IARIA

AA003981

VEHICULAR 2019

The Eighth International Conference on Advances in Vehicular Systems, Technologies and Applications

June 30 – July 4, 2019

Rome, Italy

VEHICULAR 2019 Editors

Yoshitoshi Murata, Iwate Prefectural University, Japan

Hamid Menouar, Qatar Mobility Innovations Center, Qatar University, Qatar

Printed from e-media with permission by:

Curran Associates, Inc.
57 Morehouse Lane
Red Hook, NY 12571

Some format issues inherent in the e-media version may also appear in this print version.

Printed by Curran Associates, Inc. (2019)

International Academy, Research, and Industry Association (IARIA)
412 Derby Way
Wilmington, DE 19810

Phone: (408) 893-6407
Fax: (408) 527-6351

petre@iaria.org

Additional copies of this publication are available from:

Curran Associates, Inc.
57 Morehouse Lane
Red Hook, NY 12571 USA
Phone: 845-758-0400
Fax: 845-758-2633
Email: curran@proceedings.com
Web: www.proceedings.com

Table of Contents

VEHICULAR 2019

Foreword

The Eighth International Conference on Advances in Vehicular Systems, Technologies and Applications (VEHICULAR 2019), held between June 30 – July 4, 2019 - Rome, Italy, continued the inaugural event considering the state-of-the-art technologies for information dissemination in vehicle-to-vehicle and vehicle-to-infrastructure and focusing on advances in vehicular systems, technologies and applications.

Mobility brought new dimensions to communication and networking systems, making possible new applications and services in vehicular systems. Wireless networking and communication between vehicles and with infrastructure have specific characteristics from other conventional wireless networking systems and applications (rapidly-changing topology, specific road direction of vehicle movements, etc.). These led to specific constraints and optimizations techniques; for example, power efficiency is not as important for vehicle communications as it is for traditional ad hoc networking. Additionally, vehicle applications demand strict communications performance requirements that are not present in conventional wireless networks. Services can range from time-critical safety services, traffic management, to infotainment and local advertising services. They are introducing critical and subliminal information. Subliminally delivered information, unobtrusive techniques for driver's state detection, and mitigation or regulation interfaces enlarge the spectrum of challenges in vehicular systems.

We take here the opportunity to warmly thank all the members of the VEHICULAR 2019 Technical Program Committee, as well as the numerous reviewers. The creation of such a high quality conference program would not have been possible without their involvement. We also kindly thank all the authors who dedicated much of their time and efforts to contribute to VEHICULAR 2019. We truly believe that, thanks to all these efforts, the final conference program consisted of top quality contributions.

Also, this event could not have been a reality without the support of many individuals, organizations, and sponsors. We are grateful to the members of the VEHICULAR 2019 organizing committee for their help in handling the logistics and for their work to make this professional meeting a success.

We hope that VEHICULAR 2019 was a successful international forum for the exchange of ideas and results between academia and industry and for the promotion of progress in the field of vehicular systems, technologies and applications.

We are convinced that the participants found the event useful and communications very open. We also hope that Rome provided a pleasant environment during the conference and everyone saved some time for exploring this beautiful city.

VEHICULAR 2019 Chairs

VEHICULAR Steering Committee

Markus Ullmann, Federal Office for Information Security / University of Applied Sciences Bonn-Rhine-Sieg, Germany
Carlos T. Calafate, Universitat Politècnica de València
Éric Renault, Institut Mines-Télécom | Télécom SudParis, France
Khalil El-Khatib, University of Ontario Institute of Technology - Oshawa, Canada
Manabu Tsukada, University of Tokyo, Japan

VEHICULAR Industry/Research Advisory Committee

Clément Zinoune, Renault, France
Michelle Wetterwald, HeNetBot, France
Yi Ding, US Army RDECOM-TARDEC, USA
William Whyte, Security Innovation, USA

VEHICULAR 2019

Committee

VEHICULAR Steering Committee

Markus Ullmann, Federal Office for Information Security / University of Applied Sciences Bonn-Rhine-Sieg, Germany
Carlos T. Calafate, Universitat Politècnica de València
Éric Renault, Institut Mines-Télécom | Télécom SudParis, France
Khalil El-Khatib, University of Ontario Institute of Technology - Oshawa, Canada
Manabu Tsukada, University of Tokyo, Japan

VEHICULAR Industry/Research Advisory Committee

Clément Zinoune, Renault, France
Michelle Wetterwald, HeNetBot, France
Yi Ding, US Army RDECOM-TARDEC, USA
William Whyte, Security Innovation, USA

VEHICULAR 2019 Technical Program Committee

Ali Abedi, University of Maine, USA
Eric J. Addeo, DeVry University, USA
Sufyan T. Faraj Al-Janabi, University of Anbar, Ramadi, Iraq
Michele Albano, CISTER | INESC/TEC | Polytechnic Institute of Porto, Portugal
Ignacio Alpiste, University of the West of Scotland, UK
Alexandre Armand, Renault, France
Kamran Arshad, Ajman University, United Arab Emirates
Dalila B. Megherbi, University of Massachusetts, USA
Eduard Babulak, Fort Hays State University, USA
Andrea Baiocchi, University of Roma "Sapienza", Italy
Stylianos Basagiannis, United Technologies Research Centre, USA
Marcel Baunach, Graz University of Technology | Institute for Technical Informatics, Austria
Luca Bedogni, University of Bologna, Italy
Rahim (Ray) Benekohal, University of Illinois, USA
Chafika Benzaid, University of Sciences and Technology Houari Boumediene (USTHB), Algeria
Luis Bernardo, Universidade NOVA de Lisboa, Portugal
Neila Bhouri, IFSTTAR / COSYS / GRETTIA, France
Abdelmadjid Bouabdallah, Université de Technologie de Compiègne, France
Christos Bouras, University of Patras | Computer Technology Institute & Press «Diophantus», Greece
Salah Bourennane, Ecole Centrale Marseille, France
Alexander Brummer, University of Erlangen-Nürnberg, Germany
Lubos Buzna, University of Žilina, Slovakia

Reinhard German, UniversityofErlangen-Nuremberg, Germany
Thanassis Giannetsos, DTU University, Denmark
Apostolos Gkamas, University Ecclesiastical Academy of Vella of Ioannina, Greece
Sezer Goren, Yeditepe University, Turkey
Javier Gozalvez, Universidad Miguel Hernandez de Elche, Spain
Luigi Alfredo Grieco, Politecnico di Bari, Italy
Stefanos Gritzalis, University of the Aegean, Greece
Yu Gu, Hefei University of Technology, China
Mesut Güneş, Institute for Intelligent Cooperating Systems | Otto-von-Guericke-University Magdeburg, Germany
Biao Han, National University of Defense Technology, Changsha, China
Petr Hanáček, Brno University of Technology, Czech Republic
Hong Hande, National University of Singapore, Singapore
Morteza Hashemi, Ohio State University, USA
Hiroyuki Hatano, Utsunomiya University, Japan
Ivan W. H. Ho, The Hong Kong Polytechnic University, Hong Kong
Javier Ibanez-Guzman, Renault S.A., France
Khalil ibrahimi, Ibntofail University, Kenitra, Morocco
Hocine Imine, IFSTTAR/LEPSIS, France
Mohammad Reza Jabbarpour, Islamic Azad University, Tehran-North Branch, Iran
Terje Jensen, Telenor, Norway
Han-You Jeong, Pusan National University, Korea
Yiming Ji, University of South Carolina Beaufort, USA
Magnus Jonsson, Halmstad University, Sweden
Yasin Kabalci, Omer Halisdemir University,Turkey
Arnaud Kaiser, Institut de Recherche Technologique SystemX (IRT SystemX), France
M. A. S. Kamal, Monash University Malaysia, Malaysia
Frank Kargl, Ulm University, Germany
Sokratis K. Katsikas, University of Piraeus, Greece
Abdelmajid Khelil, Landshut University, Germany
Wooseong Kim, Gachon University, Korea
Xiangjie Kong, Dalian University of Technology, China
Jerzy Konorski, Gdansk University of Technology, Poland
Dimitrios Koukopoulos, University of Patras, Greece
Zdzislaw Kowalczuk, Gdansk University of Technology, Poland
Milan Krbálek, Czech Technical University in Prague, Czech Republic
Ajith Kumar P. R., Nokia, Bangalore, India
Anh Le Tuan, Insight - NUI Galway, Ireland
Fedor Lehocki, Slovak University of Technology in Bratislava, Slovak Republic
Christian Lehsing, Harvard Medical School, USA & Technical University of Munich, Germany
Pierre Leone, University of Geneva, Switzerland
Marco Listanti, University of Roma "La Sapienza", Italy
Lianggui Liu, Zhejiang Sci-Tech University, China
Miao Liu, IBM Research, USA
Yali (Tracy) Liu, AT&T labs, Inc., USA
Seng Loke, Deakin University, Melbourne, Australia
Xuanwen Luo, Sandvik Mining, USA
Barbara M. Masini, CNR - IEIIT, Italy

Leandros Maglaras, De Montfort University, UK
Abdelhamid Mammeri, University of Ottawa, Canada
Zoubir Mammeri, IRIT - Paul Sabatier University, France
Francesca Martelli, Institute for Informatics and Telematics (IIT) - Italian National Research Council (CNR), Pisa, Italy
Chetan Belagal Math, Eindhoven University of Technology, Netherlands
Ilaria Matteucci, IIT-CNR, Italy
Natarajan Meghanathan, Jackson State University, USA
Rashid Mehmood, King Abdul Aziz University, Jeddah, Saudi Arabia
João Mendes-Moreira, LIAAD-INESC TEC | University of Porto, Portugal
Lyudmila Mihaylova, University of Sheffield, UK
Vicente Milanes, RENAULT SAS, France
Steffen Moser, Ulm University, Germany
Nils Muellner, Malardalen University, Sweden
Saeid Nahavandi, Institute for Intelligent Systems Research and Innovation (IISRI) - Deakin University, Australia
Mort Naraghi-Pour, Louisiana State University, USA
Jose E. Naranjo, INSIA | Technical University of Madrid, Spain
António J. R. Neves, University of Aveiro, Portugal
Tae (Tom) Oh, Rochester Institute of Technology, USA
Arnaldo Oliveira, Universidade de Aveiro, Portugal
Tomas Olovsson, Chalmers University of Technology, Sweden
Rachid Outbib, Aix-Marseille University, France
Markos Papageorgiou, Technical University of Crete, Greece
Al-Sakib Khan Pathan, Southeast University, Bangladesh
Xiaohong Peng, Aston University, UK
Paulo Pinto, Universidade Nova de Lisboa, Portugal
Hamid R. Rabiee, Sharif University of Technology, Iran
Ali Rafiei, University of Technology Sydney, Australia
Jacek Rak, Gdansk University of Technology, Poland
Hesham Rakha, Virginia Tech Transportation Institute, USA
Mubashir Husain Rehmani, COMSATS Institute of Information Technology, Pakistan
Éric Renault, Institut Mines-Télécom | Télécom SudParis, France
M. Elena Renda, IIT - CNR - Pisa, Italy
Jean-Pierre Richard, Centrale Lille - French "Grande Ecole", France
Martin Ring, Bosch Engineering GmbH Abstatt, Germany
Justin P. Rohrer, Naval Postgraduate School, USA
Claudio Roncoli, Aalto University, Finland
Javier Rubio-Loyola, CINVESTAV, Mexico
Larry Rudolph, TWO SIGMA, LP, USA
Marcel Rumez, University of Applied Sciences, Karlsruhe, Germany
José Santa Lozano, University of Murcia, Spain
Fareena Saqib, University of North Carolina at Charlotte, USA
Panagiotis Sarigiannidis, University of Western Macedonia, Greece
Christoph Schmittner, AIT Austrian Institute of Technology GmbH, Austria
Erwin Schoitsch, AIT Austrian Institute of Technology GmbH, Austria
Michele Segata, University of Trento, Italy
Miguel Sepulcre, Universidad Miguel Hernandez de Elche, Spain

Clément Zinoune, Renault, France
André Ventura Zúquete, University of Aveiro | IEETA research institute, Portugal

Reactive Topology-based Routing for VANETs: A New Bio-inspired Solution

An Evolutionary Algorithm-based Vehicular Reactive Protocol With Route Break Prediction Mechanism

Youcef Azzoug
Faculty of Electronic and Informatic
USTHB University, Bp 32 El Alia 16111 Bab Ezzouar
Algiers, Algeria
e-mail: yazzoug@usthb.dz

Abdelmadjid Boukra
Faculty of Electronic and Informatic
USTHB University, Bp 32 El Alia 16111 Bab Ezzouar
Algiers, Algeria
e-mail: aboukra@usthb.dz

***Abstract*—Vehicular Ad-hoc Network (VANET) routing took its early bases from Mobile Ad-hoc Network (MANET) originated routing protocols starting especially by enlarging topology-based protocols that have been conceived for MANETs, which suffered numerous modifications to make them applicable on vehicular mobility patterns. Topology-based routing is divided between proactive and reactive approaches. The latter came to fulfill the shortages of the former. In this paper, the inspiration is got from few enhanced versions of vehicular reactive topology-based routing in their two forms hop-by-hop and source routing, particularly from Ad-hoc On-demand Distance Vector Routing (AODV) and Dynamic Source Routing (DSR) originated routing protocols such as the Prior AODV and Enhanced DSR, to combine their advantages. Following this axis, an evolved Evolutionary Algorithm (EA) is exploited, in this case the Bull Optimization Algorithm (BOA): an ameliorated Genetic Algorithm (GA) incorporated with few conventional modifications inspired from the aforementioned protocols to improve the Quality of Service (QoS) routing performances. The suggested routing algorithm offers a new solution for assisting the source routing forwarding for vehicular networks.**

***Keywords*—source routing; VANETs; P-AODV; G-NET; Bull Optimization Algorithm.**

I. Introduction

The reactive routing destined for VANETs [1] has been the context of a wide research derivation, tending for improving the operational routing mechanisms, for instance the route discovery, route repair process, neighboring table optimization, beaconing, route refreshing among others. This routing model suggests two types of on-demand forwarding strategies: hop-by-hop (or next-hop) routing and source routing. For the first, each data packet tracks the destination by keeping two addresses in its header: the next-hop and destination's Identifiers (IDs). These two IDs are recorded in the routing table. This mechanism helps to perform an efficient periodic refreshing using beacons [2] to the best routes to any destination, thus an easy adaptability to quick topology changes. On the other hand, hop-by-hop forwarding may accumulate congesting routing overheads due to beaconing which is responsible for tracking neighborhood connectivity evolution. In the source routing, every data packet loads its route hops' IDs in its header to attain the destination, that does not require beaconing, the fact that packet forwarding depends mainly on intermediate hops' available full routing paths. This mechanism entails an alleviated bandwidth occupation due to elimination of beaconing-originated overheads, but such routing policy becomes less reliable when confronting to quick mobility of nodes, which characterize vehicular networks, especially for longer routes where topology changes are more frequent, hence higher number of dropped packets. AODV [3] is a typical hop-by-hop reactive protocol, whereas DSR protocol [4] is a classic routing algorithm of source routing.

This paper proposes a new bio-inspired routing algorithm to enhance the reactive routing under VANET environments based on an enhanced EA, in this case the BOA metaheuristic [5] that seeks generating better genetically-generated routing paths after the route discovery to manage data forwarding better under unpredictable vehicular mobility and drastic network density changes. The suggested solution is conceived on a multipath DSR-based route discovery procedure and completed with few AODV-evolved routing up-to-date operations.

The following context of this paper is divided as follows: Section 2 spreads the notable related contributions from both conventional and nature-inspired reactive protocols in their two types: hop-by-hop and source routing. This section clarifies the impact of bio-inspired optimization algorithms on reactive routing. Section 3 introduces the BOA. Section 4 opens out onto the suggested solution's modules. Section 5 spreads in details BVRRP's modules describing the exploited qualities of BOA metaheuristic. Finally, Section 6 regroups the conclusion and a perspective for further contributing modifications. The configuration settings adopted for the solution's codification and simulation phase are also cited in this section.

II. Related work

This section spreads some notable conventional and bio-inspired protocols in the studied context, based mostly on VANET-destined AODV-based and DSR-based protocols.

A. Conventional reactive topology-based protocols

[6] discussed numerous enhanced versions of AODV for VANETs, notably:

Improved-AODV [7] seeks route stability and less network overheads by setting two steps for route discovery and route selection phases respectively. In the first, Route Requests (RREQs) are broadcasted to the more stable nodes, i.e., the vehicles merging toward the same direction and having clause speed. In fact, more stable vehicles engender more stable route links. In the second, two selection strategies are adopted for multiple source-destination trajectories determining more stable routing paths: the first deducts stable routes by calculating the Route Expiration Time (RET), while the second calculates the route's total weight. RET represents the link having the minimum Link Expiration Time (LET) between all route links. LET is calculated using link hops' velocity, direction and communication range. Total route weight is calculated by adding up all links' weights, where each link's weight calculated by considering the velocity and direction of each link's upstream and downstream hops. It is worth noting that a predefined maximum threshold of selected neighbors following their link weight must not be surpassed so as bandwidth consumption is reduced. Finally, routes having a minimized weight and longer RET are more likely to be chosen for packet forwarding.

AODV-VANET [8] modifies both Route Request (RREQ) and Route Reply (RREP) of default AODV where each node's velocity, acceleration and direction values are added to RREQ header for route weight calculation. The latter is initialized by the source node. Route weight is the sum of all link weights passing by all hops liaising the source and destination nodes, where each current hop uses its predecessor's information to quantify the link chaining them. For RREP, the accumulated built route weight is added to its header plus the route itself and sent back straight to the source. The latter chooses among numerous received Route Replies (RREPs) the route having the minimum weight which indicates the more reliable one.

AODV-BD [9] keeps request broadcasting and data packets forwarding while route repair is being proceeded, seeking by that the dissemination of data packets to their destinations in reduced delays through fresher routes. That is, instead of launching route rediscovery, which wastes more time, data packets are rebroadcasted simultaneously with RREQs to neighboring nodes, and the one having a path to the destination takes charge of forwarding data packets to their final hop. RREP in this case is sent back to the source node when its related routed data packets reach the destination. This policy allows to shorten routing delays the fact considering the elimination of the required average time that necessitates for the reestablishment of new routes.

Prior-AODV (P-AODV) [10] seeks a higher route profitability by establishing a limited number of routes having fewer hops and fewer broken links. Thus, P-AODV restricts the number of RREQs and neighbors. In the first, for each node except the source, the number of distributed RREQs is limited to a predefined packet number threshold so that not all node's neighbors are notified. For the second, the number of neighbors is reduced, based on their distance from the current hop. These two restrictions seek next-hops that reduce the probability of link breaks. According to P-AODV, any node's surrounding neighbors are classified into two categories by considering the distance separating between the node and every neighbor: prior neighbors, which concern generally 2-hop nodes and more away, to which RREQs are sent, and overheads neighbors, which are generally 1-hop away from the current hop, whose RREQs are sent in second priority after prior zone if the aforementioned threshold allows. That is because overhead nodes share common neighbors with current hop in the prior zone. As a result, routing overheads are generally decreased.

Concerning source routing, numerous conventional protocols exist like DSR, RBVT-R [11] among others. Several DSR-modified versions have been realized, initially for MANETs and VANETs afterwards. Few cases are exposed below:

Enhanced-DSR (E-DSR) [12] adds two modifications to the default DSR which consists of: replacing the request broadcasting by multicasting to cut down overheads, and adopting a route hop limit to shorten long paths. Indeed, each node avoids distributing RREQ copies among the neighbors which are included in the in-progress request path or those sending RREQ copies to the current hop. This method helps to reduce the number of RREPs and that of control packet overheads consequently. To cope with DSR limits toward large networks, E-DSR sets a Time-To-Live (TTL) hops variable for destinations that are distant more than a customized TTL threshold from the related source node. This TTL variable is reinitialized by each node preceding the destination on expiration. Such node truncates the past hops, saves it in its route cache, reset the TTL variable, and forward the remaining path's hops until meeting the destination. This mechanism allows to alleviate data packet header and splitting data dissemination towards distant destinations into short-path forwarding sequences, hence gathering more reliable routing.

TE-DSR [13] selects route paths based on their trust evaluation, since each node saves a trust value for each of its neighbors. The degree of interactivity between these neighbors determines how these values are updated. This trust mechanism is composed of three modules. The first module includes three parts: a trust unit that involves an initializer which assigns low trust values for new unknown nodes, an upgrader for updating trust values depending on the previous experience values, and an administrator for interfacing between the previous two components and DSR protocol by storing trust information during runtime. The second is a router module which uses trust values of inter-node relationships for route evaluation and selection. The third is a monitor module which adjusts trust values for either received or missing acknowledgments. Any route discovery's received RREPs are sorted on the basis of this trust system.

For the proposed solution's context, numerous modified versions of VANET-destined DSR-originated protocols have been proposed, targeting more secured and rapid forwarding such as:

Modified DSR (M-DSR) [14] integrates a breakage prediction mechanism to the original DSR, based on tracking the Received Signal Strength Indicator (RSSI). Each node that anticipates a potential link break, when detecting the decreasing tendency of RSSI, engenders a Soon Link Breakage Warning (SLBW) packet, an equivalent of the Route Error (RERR) in default DSR, to unicast it to the source node. The latter marks with Breakage Prediction (RBP) the route having breakable link(s) with low RSSI state. Then, the source checks for another available route for data forwarding, i.e., the one that does not match with the RBP-marked routes. If it is not the case, this source node triggers a route discovery by broadcasting a Modified RREQ (MRREQ) whose header is split into RREQ header and source route header. The latter stores the RBP-marked route hops' IDs to prevent from building new paths with potential link breaks.

B. Bio-inspired reactive routing related protocols

Earlier, DSR and AODV have been ameliorated using bio-inspired metaheuristics for VANETs. Few notable works in this field are discussed below:

AODVCS [15] a Cuckoo Search (CSA)-inspired protocol for VANETs taking bases from AODV. AODVCS exploits the aggressive egg reproduction behavior of Cuckoo particles to formulate an optimized source-destination routing path among a number of established routes. This is done by the implementation the CSA [16] in optimizing the route distance toward the destination by involving levy flights, a random walk of a given cuckoo which generates new solutions, i.e., finds better nests with superior reproduction. This mechanism is applied in each node to find its next-hop which concludes to form shorten routes in term of hop count. Each RREQ is modified by adding an additional field recording the past IDs between the source and destination to use it for fitness calculation and taking hop count as the sole fitness factor. Also, an adjacency matrix is set to record all links between source and destination. This matrix is used to repair the invalid solutions that can be engendered in the Lévy flight calculation.

FA-AODV [17] a Firefly Algorithm (FA) [18] routing protocol consisting of an FA-assisted partial route selection on AODV. FA-AODV estimates the stochastic brightness of nodes for next-hop selection, where the node having more brightness, which reflects in its calculated reachability degree, is prioritized to be chosen as FA routing packet's next-hop. FA routing packets act better than RREQs and RREPs since fireflies require less messages for forming paths. As a consequence, less overheads are engendered.

Several researchers provided different nature-inspired reactive routing algorithms for VANETs. Few notable solutions are discussed below:

G-NET [19] sets a DSR-based route discovery and a GA-inspired route optimization and maintenance. G-NET mobilizes two control packets for route discovery phase: G-NET_REQUEST and G-NET_RESPONSE. GA is executed on the destination node to optimize, by genetic recombination, the collected set of chromosome request paths which forms the GA' initial population. Each chromosome is codified in a set of genomes representing route hops. The fitness formula, which evaluates each population's solution quality, considers the route latency as sole evaluation parameter since it shows the adaptation level of individuals to the vehicular environment. G-NET performs classic GA tournament selection phase, one-point crossover and mutation operators for routing optimization. The route repair procedure follows after GA mutation by eliminating the reappearing genomes within the same route so that route loops are avoided. Finally, the stopping condition is fixed to four generations to stop running the GA.

Genetic AODV (G-AODV) [20] a secure backup GA-enhanced reactive hop-by-hop topology-based routing protocol that improves participating nodes' reliability. G-AODV seeks reducing RREQs using the GA. To do so, G-AODV performs an AODV-like route discovery where the source node avoids broadcasting RREQs to all its neighbors to avoid bandwidth wastage. Instead, it utilizes the GA to discover three routes for any solicited destination. It is worth noting that G-AODV applies a terminating condition for each RREQ using three user-defined evaluation thresholds (p1, p2 and p3) that any candidate solution must surpass to stop running the GA. Finally, formed routes tend to have a reasonable cost in term of forwarding time.

III. The Bull Optimization Algorithm (BOA)

Oguz Findik suggested the BOA, an enhanced version of the GA for continuous optimization problems [5]. BOA is a GA-based metaheuristic that implements the genetic crossover and mutation operators. BOA seeks surpassing GA's research depth limitations which reside in two parts, in this case: the decreasing stochastic search abilities when selecting the best individuals since early stages, and the quick convergence to local optima caused by the random selection of partial solutions having good fitness. Thus, BOA eliminates the selection phase which is usually behind the aforementioned weaknesses. Also, BOA increases the mutation percentage which imposes an influent randomization impact so that the convergence to local minima induced by the GA's low mutation rate is avoided. BOA keeps use of initial population during all process to exploit worse solutions to generate better global optima. To summarize, BOA is expected to offer less computation complexity than GA. The BOA cycle is illustrated in Figure.1.

IV. Our suggestion: BOA-assisted Vehicular Reactive Routing Protocol (BVRRP)

Numerous conventional enhanced protocols of AODV and DSR gave effective ideas on optimizing beaconing process, shortening the route discovery-repair sequence, alleviating request packet headers and so on. This helps to reduce forwarding delays and overheads and raise up packet delivery ratio. Meanwhile, it is noticed that reactive topology-based routing suffers from difficulties to maintain longer routing paths due to topology changes and high routing overheads generated by the beaconing. Also, such protocols are solution-orientated, since each protocol tries to solve a particular problem. For instance, the case of P-AODV which focuses on request broadcasting optimization or the case of M-DSR which concentrates on route break prediction optimization.

BVRRP protocol is destined for source routing assistance. Its purpose is to use the collected routes from the route discovery as an initial population for proceeding a BOA-based stochastic recombination of request paths. This protocol combines few foundations suggested in few DSR-modified and hop-by-hop routing algorithms with route optimization inspired from G-NET.

The strategy of BVRRP is resumed in the principles below:

- BVRRP is a source routing protocols, the case of multi-path DSR [21] and DSR-based protocols.
- BVRRP builds multiple link-disjoint routes for the same route demand, where routes must share one or numerous nodes, which is mandatory for BOA's genetic operators.

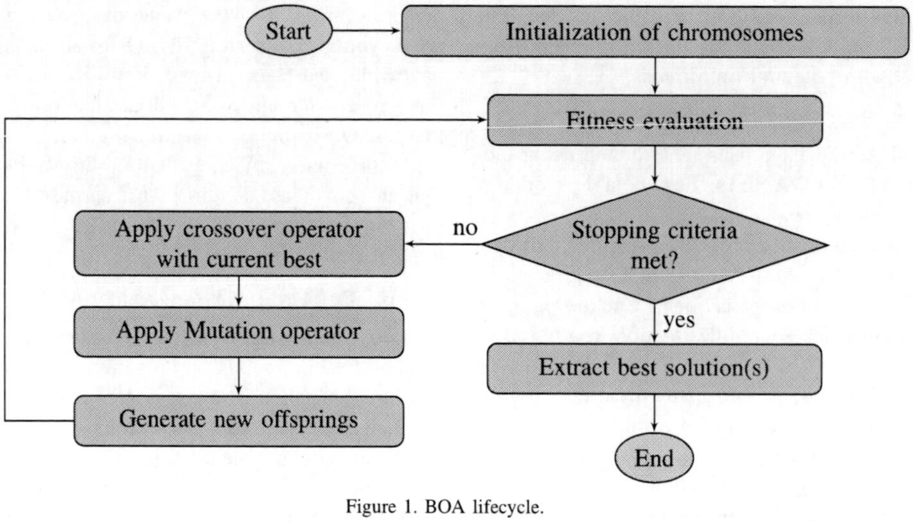

Figure 1. BOA lifecycle.

- The route discovery in BVRRP is based on a default request broadcasting as performed in AOMDV [22] and multipath DSR except its duration is adapted seen the amount of collected requests.
- BVRRP adopts a restricted beaconing for refreshing both neighborhood and topology links information which is required for BOA application and route break prediction mechanism.
- BVRRP implements two fitness optimization formulae: the first considers the parameters that favors limited lifetime routes providing quick routing, whereas the second offers longer-living routes to fit with recovery routing conditions constrained with considerable structural topology changes.
- BOA's first fitness formula, which is conceived for ordinary forwarding, involves three parameters for its calculation, in this case: the request path's hop count, average neighborhood longevity of request path's hops and request delay. The generated routes prioritize short route length which entails a quicker data forwarding and avoids routing circumstances under unmanageable mobility changes. Thus, the probability of switching to route recovery mode is reduced.
- The second fitness formula is conceived for route break recovery and implies three parameters: request delay, average number of route breaks of request path's hops and average RSSI of request path's hops. The generated routes are tendentious to have a longer longevity rather than a shorter path length so that more route reliability is gathered in order to adapt to route recovery forwarding conditions.
- BVRRP shortens BOA runtime on the destination by advancing the stopping criteria, when the number of collected routes is under a predefined route number threshold, since the initial population is not filled enough to extend for a deeper optimization reach.
- Furthermore, BVRRP adjusts BOA runtime depending on a predefined average route hop count threshold. An extension of stopping criteria is set for longer routes while a restriction is reserved for shorter ones. This serves a better research profitability of BOA.
- Recovery forwarding: BVRRP runs route repair phase with data packet forwarding in a parallel paradigm when the regular forwarding fails. Thus, the best available BOA-generated path is used for recovery forwarding while RERR packet is sent back to the source by backward path for route break notification and routing cache updates.
- BVRRP sets a link break prediction mechanism performed on each BOA-built route notifying all its hops and their limited neighborhood area about the state indicators of potential breakable links. This is done by checking up-to-date RSSI values collected by hello packets on each hop.

V. Contribution details

BVRRP is an intelligent reactive routing protocol implementing the BOA for route optimization, parallel route repair with recovery data fowrding, and route break prediction system as detail in this section, where the main modules for this protocol are described detailing the major set of modifications, including the positioning of BOA in the routing proceedings:

A. Route discovery phase:

BVRRP's route discovery is an approach founded from the general principle of multipath routing, implemented in AOMDV and multipath DSR, based on collecting numerous routes for the same destination using source route discovery. Such routes can share one or numerous hops to cope with the requirements of BOA's genetic operators. Whereas, paths are link-disjoint since sharing route links degrades the stochastic impact of crossover and mutation operators to produce better paths. Every RREQ saves all passed nodes to form a request path. All collected paths from the received RREQs in the destination node will constitute the BOA's initial population. The routing discovery duration is limited by a Request Time-To-Live (RTTL) which marks how many RREQs are allowed to be received by the destination. The RTTL can be:

- Number of RREQs threshold (Nbr_{RREQ}): marks the maximum allowed number of received RREQs, i.e., represents the size of BOA's initial population.
- A discovery duration threshold ($Delay_{RD}$): this factor set an expiration time for the route discovery and it is enabled in case of the number of received RREQs is still under the predefined Nbr_{RREQ}.

BVRRP-conceived RREQ adds further fields to gather information required for checking link-disjointless property, evaluating route links quality, and regrouping necessary information for BOA's fitness evaluation mentioned in the previous section. Each RREQ saves, for each traversed hop, an entry including its average neighborhood longevity and its number of downstream link breaks. A request next-hop list is recorded as well containing each recorded node with its next hop. That will serve for BOA's mutation operator. Whereas, each RREP records one BOA-generated routing path. BVRRP's RREQ and RREP packet structure are illustrated in Table. I and Table. II, respectively.

Each hop receiving a RREQ performs the below-mentioned control operations to decide either accepting or rejecting it:

- If the current hop's ID is already recorded in the RREQ header's in-progress request route then this RREQ is discarded, since a loop over current hop is detected.
- Else, if the pair $\{$source ID, RREQ ID$\}$ is already passed by the current hop, then the received RREQ has been already treated, so this RREQ is discarded as a result.
- Else, if any of the links recorded in the received RREQ have been already frequented from an earlier RREQ by the current node for the same destination, then the received RREQ is discarded, since the route link-disjointless property is not guaranteed in this case.
- Else, if the current hop's ID is the destination's ID then if the route discovery time limit is not expired yet, by checking both Nbr_{RREQ} and $Delay_{RREQ}$ thresholds, then the build request path is added BOA's initial population, otherwise the destination rejects all new arriving RREQs for the corresponding source-destination route discovery and the BOA routing optimization is started.
- Otherwise, the received RREQ is the first copy of its source-destination entry, so it's ID is appended to the in-progress request route if its route hop count is under the predefined hop count limit (HTTL) threshold which marks the maximum allowed length for request chromosomes of BOA's initial population.

Furthermore, each valid received RREQ (accepted request) gets to add the following information when hosted by each intermediary node:

- Updating the average RSSI of all passed hops up to the host hop.
- Updating the average number of route link breaks of all passed hops including the current hop.
- Updating the average request path's neighborhood lifespan with the current hop's average neighbors' lifetime.
- Recording the last and current hop in the request next-hop list. As mentioned, this serves performing the BOA's mutation operator.
- Recording the accumulated delay since the sent of the RREQ from source node.

B. Restricted Beaconing

A restricted beaconing is executed to serve the BOA application and route break estimation in which hello notification messages are sent periodically to 1-hop neighboring nodes for:

- Recording reachable 1-hop neighbors: each node adds a new node on its neighbors table and records the timestamp when it was added. This permits to quantify the lifespan of the liaison between the two hops which will serve for BOA main route fitness evaluation. In our opinion, it is estimated that every vehicle having a longer average connection with its neighbors is supposed to perform more reliable routing.
- Recording timestamp and value of hello-collected RSSI for each neighbor. This measure serves for route break prediction mechanism, offering more recent information about each route links's connectivity state.
- Updating the neighbor table and what entails as changes in routing cache and request neighborhood list. Such updates are mandatory to perform BOA's fitness calculation.

The periodic interval that separates two hello packets' broadcasting is dynamically adapted depending on the node's solicitation degree which refers to the its participation level to data forwarding in the network. This indicator is approached following the below factors in any instant *t*:

- Number of active neighbors ($Neigh_{Count}$).
- Number of active request entries ($Active_{RREQs}$).
- Unreachability degree, measured by the historic number of route link breaks as downstream hop ($RouteBrks_{Count}$).

Any node's reachability value ($Reach_{Level}$) is calculated using (1).

$$Reach_{Level} = (0.3 \times Neigh_{Count}) + (0.6\times Active_{RREQs}) + (0.1 \times RouteBrks_{Count}) \quad (1)$$

A solicitation threshold ($Solicitation_{Thresh}$) is predefined to decide whether accelerating or slowing down the beaconing frequency, where any node having a higher reachability value than $Solicitation_{Thresh}$ obliges more hello updates to avoid quick data aging (2).

$$Reach_{Level} \leq Solicitation_{Thresh} \quad (2)$$

Table. III presents the structure of the proposed neighbor table set on each node.

C. Application of BOA

The BOA optimization is performed in the destination node straight after expiration of the RTTL in case no notification has been received on the source by any intermediate node that recognizes the destination. The collected routes constitute the initial population of BOA which passes a limited number of iterations to produce few optimized routes. This iteration limit is adjusted basing on two factors:

- *Average Hop Count of initial population's individuals (AvgHC)*: based on request chromosomes's length since longer routes offer more recombination alternatives. This

TABLE I. BVRRP'S RREQ PACKET STRUCTURE.

Packet type	Request ID (Seq number)	Source ID	Destination ID
Path hops IDs	Hop Count	Total delay	Neighborhood list
Avg Nbr Route breaks	Avg Recv signal strength	Next-hop list	Topology links list

TABLE II. BVRRP's RREP PACKET STRUCTURE.

Packet type	Source ID	Destination ID	Route type (Main/Recovery)	Route's IDs

TABLE III. NEIGHBOR LIST'S ENTRY FIELDS.

1-hop neighbor ID	Neighbor record time	Neighbor's hello RSSI	RSSI timestamp	Neighbor Delay

parameter decides either reducing or extending BOA lifetime, eliminating BOA for main route (restricting BOA application to recovery routes), or setting a shortened BOA for either main or recovery routes optimization. Thus, a predefined hop count threshold ($Avg_{Hopcount}$) is set to decide how many generations BOA performs for both main route and recovery procedures.
- *Minimum number of RREQ routes for initial population ($MinNbr_{RREQ}$)*: decides the stopping criteria of BOA for both main route and recovery routes optimization. If the number of collected RREQ routes is under $MinNbr_{RREQ}$ within an expired $Delay_{RREQ}$ then the number of BOA generations is shortened for both main and recovery optimization procedures.

The full BOA cycle performed in the destination node is detailed below:

- *Solution codification*: Each chromosome represents a candidate source-destination route, where the genomes represent route hops. It is worth noting that BOA chromosomes do not share the same dimension, since routes can have different hop count. Two hop count limits for main path (HC_{Main}) and recovery paths ($HC_{Recovery}$) are set respectively to avoid extended routes, while the BOA genetic operators, i.e., crossover and mutation are customized to avoid any dysfunction that may entail erroneous paths. It is worth noting that HC_{Main} is shorter than $HC_{Recovery}$ to meet the above-described routing constraints.
- *Initialization step*: The initial population regroups all established node-shared paths collected from received RREQs that liaise the source and destination nodes. Individuals differ in term of hop count; hence initial population's chromosomes have variable length which forces few adjustments in crossover operator. Noting that initial chromosomes can be longer than HC_{Main} and $HC_{Recovery}$ to open the possibility for generating better offspring paths after crossover and mutation recombination. It is worth to mention that BOA's initial population is kept during all performed BOA generations.
- *Fitness functions*: each initial chromosome is evaluated with two fitness equations used for main and recovery procedures of BOA algorithm respectively. The $Fitness_{routing\_path}$ calculated for the main BOA route fitness and $Fitness_{recovery\_paths}$ calculated for recovery BOA route fitness.
- *Crossover operator*: BOA's crossover operator intervenes to combine the best parts of numerous chromosomes by the permutation of route pair portions that share one or several nodes. For BVRRP, two-point crossover is performed in all BOA generations by associating the best chromosome with every member of the initial population including the partial-best chromosome. Meanwhile, since the crossover operator in BVRRP solution is variable due to the dependency from request paths' genomes where it differs from a chromosome pair to another, depending on the position of common permutation points, either one-point crossover is switched in case there is one common node between chromosomes pair or no crossover if there are no common genomes.
- *Mutation operator*: BOA's mutation operator intervenes for tracking global-varied solutions. It consists of replacing a randomly-selected genome of a candidate solution with a random hop provided that it belongs to the next-hop list of the replaced genome's predecessor. A low mutation probability (M_{prob}) is set for BVRRP solution (M_{prob}= 0.02) to avoid high variety impact which risks converging to lower-quality solutions. Noting that a two-point mutation is applied for BVRRP and performed on initial population's individuals including the partial-best chromosome.
- *Stopping criteria*: two stopping criteria are set for BOA. The first is static and fixed to a limited number of iterations for both main and recovery procedures. In our case, it varies between 03 and 06 generations depending on the above-mentioned iteration limit factors, i.e., adapting to both $MinNbr_{RREQ}$ and AvgHC thresholds. The second is dynamic and considers a predefined BOA-progressing factor of the global best, set at an improvement of 60% of the initial best chromosome, that must be reached to stop the BOA for both the main path's procedure and the average fitness of best five (05) solutions for the recovery paths' procedure.
- *Solutions' extraction*: the best found path from main route algorithm, i.e., having the lowest fitness value, is selected for regular forwarding, while the best five (05) solutions from recovery routes algorithm, which have the five lowest fitness values ($Min\{Fitness_{recovery\_paths}\}$), are preserved for recovering any possible route break.

The fitness function of a given route **R** to a destination **D** used for post route discovery forwarding is given in (3):

$$Fitness_{routing\_path} = (p_{\alpha 1} \times path_{delay}) + (p_{\alpha 2} \times path_{hopcount}) - (p_{\alpha 3} \times path_{rss}) \quad (3)$$

Where:

- $p_{\alpha 1}$, $p_{\alpha 2}$ and $p_{\alpha 3}$ are user defined parameters which are set in a way that favors quick forwarding over route reliability.
- $path_{delay}$ is the total delay consumed between the source and destination nodes.
- $path_{hopcount}$ is the number of chromosomes's passed nodes (hops).
- $path_{rss}$ is the average RSSI of received RREQs of all path's hops.

The main fitness parameters are set as: $p_{\alpha 1}$= 0.45, $p_{\alpha 2}$= 0.3 and $p_{\alpha 3}$= 0.25.

The fitness function used for route recovery forwarding is given in (4):

$$Fitness_{recovery\_paths} = (p_{\beta 1} \times path_{delay}) + (p_{\beta 2} \times path_{nrb}) - (p_{\beta 3} \times path_{neighborhood}) - (p_{\beta 4} \times path_{rss}) \quad (4)$$

Where:

- $p_{\beta 1}$, $p_{\beta 2}$, $p_{\beta 3}$ and $p_{\beta 4}$ are user defined parameters whose values advantage extending route longevity with keeping reasonable routing delays.
- $path_{nrb}$ is the average number of route breaks that includes the average route breaks number of chromosome hops.
- $path_{neighborhood}$ is the average neighborhood longevity of all route nodes' neighbors.

The recovery fitness parameters are set as: $p_{\beta 1}$= 0.15, $p_{\beta 2}$= 0.3, $p_{\beta 3}$= 0.3 and $p_{\beta 4}$= 0.25.

A RREP is created to record each BOA-optimized path in its header including the path links and extracted partial routes'with its fitness values, and then returned to the source node. This serves keeping the best routes and provisioning intermediate nodes having buffered data packets for the same destination. The destination node empties its structures related to the BOA optimization process for the corresponding source-destination entry in the next-hop table, Request route neighborhood table, Request topology links table and BOA table.

Each intermediary hop from the BOA-built route which receives a RREP performs the following operations:

- Updating the routing table with extracted both main and recovery RREP's sub-paths that liaise the current hop to the destination.
- Checking if there are buffered data packets for the destination and the node preceding the destination. If so, these data packets are forwarded using the main routing path and then, route discovery is avoided.
- Discharging the routing data structures related to the concluded route discovery of the source-destination entry, in this case the seen links list and the seen requests list.

The full BOA's pseudo-code implemented for route optimization is presented in Algorithm. 1.

D. Route maintenance and route break prediction

BVRRP anticipates with a route recovery strategy to face any unexpected route break and keeps data packet forwarding while performing route repair tasks. BVRRP's route maintenance phase is composed from two major mechanisms:

- *Parallel route repair and recovery forwarding system*: this module is triggered when a route break event is stated. It performs simultaneously a route repair process while keeping data forwarding. Initially, a RERR copy is generated from the node detecting the link break generally the node preceding the broken link (upstream node), and then sent back to the source node. Each intermediary hop receiving this RERR performs the necessary updates to the destination entry of its routing cache. Second, the downstream node of broken link keeps forwarding both blocked and buffered data packets using its available recovery path. In the meantime, a new route discovery is triggered by the source after being notified by the RERR.
- *Route break prediction system*: this module is the post-BOA phase, when a prediction packet (PRED) is sent back to the source just after the RREP by the reverse path. In a first step, PRED notifies built-BOA path's hops of any changes in RSSI state to detect breakable route links of this path, so that every hop that hosts this PRED updates breakable link list and marks all routes in routing cache having breakable links to restrict their lifespan, as a result, route breaks are avoided. Also, the stored breakable links have a limited lifetime and expire once the established BOA path is broken or expired. Second measure, each passed hop belonging to the built-BOA route notifies, with a PRED copy, the nodes that belongs to its three-hop radius area. This measure is set to reduce the impact of breakable links in close network area of established route and anticipate recovery forwarding which shortens routing delays. A route break equation is defined to estimate route links' lifetime as coded in the route break estimation pseudo-code presented in Algorithm. 2.

VI. Conclusion and Future Work

In this paper, an EA-assisted V2V reactive routing protocol for VANETs is suggested named the BOA-based Vehicular Reactive Routing Protocol (BVRRP) that makes use of BOA metaheuristic to generate better genetically-optimized routes for both regular and recovery data forwarding. It is noticed through this proposition the possibility of spreading EAs family in route optimization for vehicular networks and its effective implementation with other modifications that touches the main functional parts of a typical ad-hoc reactive routing protocol such as route maintenance and route break prediction. This work is opened for other modifications such the implementation of nature-inspired metaphor-based algorithms, the passage to Vehicle-To-Infrastructure (V2I) routing through cloud computing technologies, or the combination with geography-assisted routing.

Algorithm 1 BOA pseudo-code of BVRRP

```
Define Src: source node;
Define Dest: destination node;
Define constant main_threshold: loop limit for main route optimization;
Define constant rec_threshold: loop limit for recovery routes optimization;
Define constant main_optimum: minimum routing path fitness value for post route discovery forwarding;
Define constant rec_optimum: minimum routing path fitness value for post route break forwarding;
Define constant cross_prob: crossover probability;
Define constant mut_prob: mutation probability;
Declare cross_offspring: post-crossover offspring variable;
Declare mut_offspring: post-mutation offspring variable;
Declare fittest_recovery_list: vector for best 05 recovery paths;
Declare current_mbest: current best main path variable;
                                                          ▷ intialize populations and global bests
Set main_population = Initialize (initial_population);    ▷ a copy of initial population for main algorithm
Set recovery_population = Initialize (initial_population);    ▷ a copy of initial population for recovery algorithm
Set fittest_main = Best_Fitness (main_population);    ▷ get best initial chromosome for main population
Set fittest_recovery = Best_Fitness (recovery_population);    ▷ get best initial chromosome for recovery population
Set main_iteration = 1, rec_iteration = 1;
while main_iteration < main_threshold and fittest_main > main_optimum do    ▷ perform BOA for regular forwarding
    current_mbest = extract_best (main_population);
    for each individual k in main_population do
        cross_offspring = Crossover_operator (k, cross_prob);
        fittest_main = Fitness (cross_offspring);
        if fittest_main < Fitness (current_mbest) then
            Update_best (fittest_main);
        mut_offspring = Mutation_operator (k, mut_prob);
        fittest_main = Fitness (mut_offspring);
        if fittest_main < Fitness (current_mbest) then
            Update_best (fittest_main);
    main_iteration++;
if main_iteration ≥ main_threshold or fittest_main ≤ main_optimum then    ▷ select best BOA main path
    Set best_main_route = current_mbest;
    Load_RREP (best_main_route);    ▷ load new path in a RREP
    Unicast_RREP (Src, best_main_route);    ▷ send RREP to the source node
    Route_break_prediction (Src, Dest, best_main_route);    ▷ trigger route breaks prediction for sent path
while rec_iteration < rec_threshold and fittest_recovery > rec_optimum) do    ▷ perform BOA for recovery forwarding
    Set current_rbest = extract_best (recovery_population);
    for each individual j in recovery_population do
        cross_offspring = crossover_recovery_operator (j, cross_prob);
        fittest_recovery = Fitness (cross_offspring);
        if fittest_recovery < Fitness (current_rbest) then
            Update_bests (fittest_recovery_list, cross_offspring);    ▷ update best recovery routes list after crossover
        mut_offspring = mutation_recovery_operator (j, mut_prob);
        fittest_recovery = Fitness (mut_offspring);
        if fittest_recovery < Fitness (current_rbest) then
            Update_bests (fittest_recovery_list, mut_offspring);    ▷ update best recovery routes list after mutation
    rec_iteration++;
if rec_iteration ≥ rec_threshold or fittest_recovery ≤ rec_optimum) then    ▷ extract best 05 BOA recovery paths
    Set recovery_routes_list = Extract (fittest_recovery_list);    ▷ order best 05 recovery routes
    for each recovery_path rp in recovery_routes_list do
        Load_RREP (rp);    ▷ load new path in a RREP
        Unicast_RREP (Src, rp);    ▷ send RREP to source node
        Route_break_prediction (Src, Dest, rp);    ▷ trigger route breaks prediction for sent path
```

Algorithm 2 BVRRP break prediction procedure of a post-BOA route for Current hop N

Define const $HSignThresh$: packet RSSI threshold;
Define const Exp_param: expiration parameter $\in$ [0,1];
Define const Ext_param: extension parameter $\in$ [0,1]; with $Ext\_param > Exp\_param$;
Define Boa_route: routing path loaded in the PRED;
Define Src: source node;
Define $Dest$: destination node;
for each passed hop $S \in$ post-BOA path positioned after N **do**
    Record HT: last received hello signal timestamp of the link [S-1, S]; ▷ (S-1) is the previous hop of S
    Record PT: current received prediction signal timestamp of [S-1, S];
    Record HSs: last received hello signal strength of [S-1, S];
    Record PSs: current received prediction signal strength of [S-1, S];
    if $HSs > PSs$ **then** ▷ (S-1) is less reachable by S
        if $PSs < HSignThresh$ **then** ▷ (S-1) is out of transmission range of S
            Remove_link ([S-1, S], $topology\_link\_list$); ▷ delete [S-1, S] from the topology links list
            Remove_Routes ([S-1, S], $route\_cache$); ▷ delete all paths having [S-1, S]
        else
            Set $time\_exp = (PT - HT) \times (HSs - PSs) \times Exp\_param$;
            Expire_link ([S-1, S], $time\_exp$); ▷ [S-1, S] link will expire to $time\_exp$
            Update_Routes ($route\_cache$, [S-1, S], $time\_exp$); ▷ shorten paths lifespan having [S-1, S] to $time\_exp$
    else ▷ [S-1, S] is more reliable than when on path creation timestamp
        Set $time\_ext = (PT - HT) \times (PSs - HSs) \times Ext\_param$;
        Extend_link ([S-1, S], $time\_ext$); ▷ [S-1, S] lifespan is extended to $time_ext$
        Update_Routes ($route\_cache$, [S-1, S], $time\_ext$); ▷ extend paths lifespan having [S-1, S] to $time\_ext$
    Notify_Neighborhood (); ▷ notify 3-hop neighbors of evaluated path
    if breakable_link [S-1, S] and $buffer\_cache$ not empty **then** ▷ start new route discovery for $Dest$
        Route_Discovery ($Dest$);
    if (S-1) is Src **then** ▷ end route prediction process
        Stop_Process ();

It is worth noting that the BVRRP is programmed using the GloMoSim simulator in C++ and set for comparison with P-AODV and G-NET on three QoS metrics, namely: the Average End-To-End Delay (AE2ED), Packet Delivery Ratio (PDR) and the Normalizd Routing Load (NRL). The BVRRP version of *"VEHICULAR 2019"* conference is reduced to only theoretical solution for conference oral lecture while the simulation results will be attached to the BVRRP manuscript for proceeding indexing.

REFERENCES

[1] P. K. Pagadala and N. M. S. Kumar, "A survey on Topology based Reactive Routing Protocols in Vanets," *Global Journal of Computer Science and Technology: ENetwork, Web & Security*, vol. 18, no. 4, Dec. 2018. [Online]. Available: https://computerresearch.org/index.php/computer/article/view/1765

[2] K. Z. Ghafoor, J. Lloret, K. A. Bakar, A. S. Sadiq, and S. A. B. Mussa, "Beaconing Approaches in Vehicular Ad Hoc Networks: A Survey," *Wireless Personal Communications*, vol. 73, no. 3, pp. 885–912, Dec. 2013.

[3] C. E. Perkins and E. M. Royer, "Ad-hoc On-Demand Distance Vector Routing," University of California, Tech. Rep. DEC-TR-506, Aug. 2003.

[4] D. B. Johnson and D. A. Maltz, *Dynamic Source Routing in Ad Hoc Wireless Networks*. Springer, 1996, pp. 153–181.

[5] O. Findik, "Bull optimization algorithm based on genetic operators for continuous optimization problems," *Turkish Journal of Electrical Engineering & Computer Sciences*, vol. 23, pp. 2225–2239, Sep. 2013.

[6] A. Joshi, P. Sirola, and K. C. Purohit, "Comparative Study of Enhanced AODV Routing Protocols in VANET," *International Journal of Computer Applications*, vol. 96, no. 18, pp. 22–27, 2014.

[7] B. Ding, Z. Chen, Y. Wang, and H. Yu, "An Improved AODV Routing Protocol for VANETs," in *IEEE 2011 International Conference on Wireless Communications and Signal Processing (WCSP)*, Nanjing, China, Nov. 2011.

[8] H. Guo, W. C. Wong, F. B. A. Thani, and Y. Wu, "An Optimized Routing Protocol for Vehicular Ad Hoc Networks," in *TENCON 2010 - 2010 IEEE Region 10 Conference*, Fukuoka, Japan, Nov. 2010.

[9] B. Li, Y. Liu, and G. Chu, "Improved AODV Routing Protocol for Vehicular Ad hoc Network," in 3rd *International Conference on Advanced Computer Theor and Engineering(CTE)*, Chengdu, China, Aug. 2010.

[10] O. Abedi, R. Barangi, and M. A. Azgomi, "Improving route stability and overhead of the AODV routing protocol and makeing it usable for VANETs," in 29th *IEEE International Conference on Distributed Computing Systems Workshops*, Montreal, QC, Canada, Jun. 2009.

[11] J. Nzouonta, N. Rajgure, G. Wang, and C. Borcea, "Vanet Routing on City Roads Using Real-Time Vehicular Taffic Information," *IEEE Transactions On Vehicular Technology*, vol. 58, no. 7, pp. 3609–3626, Sep. 2009.

[12] S. Sultana, S. Begum, N. Tara, and A. R. Chowdhury, "Enhanced-DSR: A New Approach to Improve Performance of DSR," *International Journal of Computer Science and Information Technology*, vol. 2, no. 2, pp. 113–123, Apr. 2010.

[13] N. Bhalaji, A. R. Sivaramkrishnan, S. Banerjee, V. Sundar, and A. Shanmugam, "Trust Enhanced Dynamic Source Routing Protocol for Adhoc Networks," *World Academy of Science, Engineering and Technology*, vol. 49, pp. 1074–1079, 2009.

[14] K. Zahedi, Y. Zahedi, and A. S. Ismail, "Enhancing the Performance of DSR Routing Protocol Using Link Breakage Prediction in Vehicular Ad Hoc Network," *International Journal of Computer Networks and Communications Security*, vol. 1, no. 1, pp. 7–14, Jun. 2013.

[15] A. Kout, S. Labed, S. Chikhi, and E. B. Bourennane, "AODVCS, a new bio-inspired routing protocol based on cuckoo search algorithm for mobile ad hoc networks," *Wireless Networks - Springer*, vol. 24, no. 7, pp. 2509–2519, Oct. 2018.

[16] X.-S. Yang and S. Deb, "Cuckoo Search via Lévy flights," in *2009 World Congress on Nature & Biologically Inspired Computing (NaBIC)*, Coimbatore, India, Dec. 2009.

[17] M. B. B. Anbu Malar *et al.*, "Firefly Algorithm For Optical Path Detection In Ad Hoc On-Demand Distance Vector (AODV)," *International Journal of Advance Research in Science and Engineering*, vol. 6, no. 11, pp. 2039–2047, Nov. 2017.

[18] X.-S. Yang, "Firefly Algorithm, Lévy Flights and Global Optimization," in *Research and Development in Intelligent Systems XXVI*, Oct. 2009, pp. 209–218.

[19] E. C. G. Wille, H. I. D. Monego, B. V. Coutinho, and G. G. Basilio, "Routing Protocols for VANETs: An Approach based on Genetic Algorithms," *KSII Transactions on Internet and Information Systems*, vol. 10, no. 2, pp. 542–558, Feb. 2016.

[20] C. Garg and B. Wadhwa, "G-AODV: A Novel Approach to Improve AODV by Using Genetic Algorithm in VANET," *International Journal of Innovative Research in Computer and Communication Engineering*, vol. 4, no. 6, Jun. 2016.

[21] A. Boursier, S. Dahlen, J. Marie-Francoise, T. S. Marin, and S. Nethi, "Multipath DSR protocol for ad hoc network," Aalborg University, Institute of Electronic Systems, Tech. Rep., Dec. 2004.

[22] M. K. Marina and S. R. Das, "Ad hoc on-demand multipath distance vector routing," *Wireless Communications and Mobile Computing*, vol. 6, pp. 969–988, 2006.

A Blockchain Approach towards Cargo Sharing in Last Mile Logistics

Johannes Kretzschmar

Institute for Mathematics and
Computer Science
Friedrich Schiller University
Jena, Germany
Email: johannes.kretzschmar@uni-jena.de

Felix Eckardt

Institute for Mathematics and
Computer Science
Friedrich Schiller University
Jena, Germany
Email: felix.eckardt@uni-jena.de

Abstract—**Sharing platforms for freight orders are widely used in logistics today. Through the cross-company sharing, resources can be saved ecologically and economically in the long term. In the field of urban logistics however, such platforms are difficult to implement because the operational scope and profit margin are much smaller. This paper presents a method to implement cargo sharing in the Last Mile context using a peer-to-peer (P2P) network solution based on blockchains and smart contracts. Blockchains allow a secure, verified and consensus-based exchange of information while smart contracts ensure the organization of matchmaking, reliable contracting and order execution. The tight linkage of both technologies to cryptocurrencies and existing platforms also ensures the possibility of easy financial balancing and protecting general conditions.**

Keywords–Blockchain; Last Mile; Logistics; Multiuse.

I. Introduction

The field of logistics is particularly affected by changes over the last years, especially the so-called Last Mile. This term refers to the last step of long supply and delivery chains, which are characterized by short-distance transport processes from a central starting to multiple endpoints or vice versa. There are some critical aspects that require a rethinking and redesign of these kinds of logistics:

The ongoing demographic change and shifting of medial reception from print to online disrupt the business model of media distribution. Subscriptions are a mainstay of the newspaper industry and the omission of these over the last decade critically affects the economic efficiency of distribution tours. While the delivery of newspapers is becoming increasingly unprofitable, the demand for deliveries and services on the last mile is increasing. The displacement towards online shopping and service platforms requires a more extensive infrastructure of Courier, Express, and Package (CEP) transport providers for everyday applications like food retailers or other local suppliers like pharmacies. Finally, there is a strong movement towards environmental protection, especially pollutant and noise emissions. Last mile processes are mostly embedded in urban areas, which are more and more protected by constrained delivery areas or times as well as driving restrictions or even no driving areas for specific vehicles. There is a foreseeable development towards using vehicles without combustion engines. Alternatives hereby are drives from renewable energies like electro mobile or hybrid-driven cars or even new kinds like pedelecs, which have to be evaluated and taken into account for future tour planning.

The research project Smart Distribution Logistik (SDL)[1], on which this paper is based, investigates how newspaper publishers and distributors can address the issues mentioned above. The objective is to determine under which conditions and with which strategies delivery processes can be established long-term cost-efficiently in the future. There are several entangled approaches like open up new business fields, adapting tours, re-slicing delivery areas, reconsidering the vehicle fleet or implementing new logistics concepts like hubs. All these in-house approaches are of course limited. Studies like [2] have shown that the future of last mile logistics lies in cross-company optimization. The goal is to utilize resources, such as electrically driven vehicles as comprehensively as possible to increase cost efficiency.

In the next Section II, fundamental terms and currently used technologies will be introduced to illustrate the context of this work. In the following Section III , a communication platform will be presented, which matches supplies and demands in services and resources of companies to find possible synergies. In the course of discussing the functionality, we present a basic data model for making inquiries in Section III-A and discuss the implementation into a smart contract operating on a blockchain in Section III-B. Further, we present in Section III-C an architecture concept of how to combine a blockchain client with logistic data sources. We conclude this paper in Section IV with discussing several key factors, which such a platform has to fulfill in the context of Last Mile logistics and show how these can be achieved with blockchain technology. While working on the demonstrator, several modeling decisions were made, which are a basis for future work presented in the final Section V.

II. State of the Art in Cross-Company Logistics

With the advent and development of digital technologies, data and networks, a whole range of new products and services have emerged. The systematic cross-linking of hard- and software revolutionized entire supply chains based on monetizable data and data processing. This led to a disruption of whole market segments under the concept of platform economy. These platforms are characterized by their functionality as a central digital marketplace that matches supply and demand and thus brings together various stakeholders [3]. Contrary to the participants of the traditional market, platform operators do not have to bring any actual resources in the segment but only offer the service of matchmaking.

A. Logistics Platform Economy

Logistics has also proven to be a successful application for sharing platforms. The term 4th Party Logistics (4PL) was coined at an early stage in this field to describe the coordination and mediation of logistics service providers and infrastructure. The involved stakeholders are classified into the lower layers 3PL for partners with assets in supply chain management, 2PL for service providers in transport and storage handling and 1PL for manufacturing companies without logistic resources. The different categories may overlap and play various roles, as described in detail in [4]. Of course, the concept of 4PL is closely linked to platform economy, as early direct technological implementations of a marketplace for logistics services, such as [5], show.

Today, there is a large number of commercial solutions in this sector, which are widely used. These arise either from existing logistic service providers who expand their business model in the field of logistics IT to include a corresponding freight exchange or marketplace functionality. These providers like *TimoCom*, *Trans.eu* or *Teleroute* mostly originate from long-haul transports and offer a freight exchange platform specialized in this field. Despite, today there also exist a lot of start-ups in this area, offering services, which target a more private sector like *Shiply*, *uShip* or *Saloodo*.

As described in [3] sharing economy influences traditional markets as well as logistics in particular two ways. On one hand, the market opens to private individuals offering resources or services. On the other, sharing platforms quickly tend to play significant and market-determining roles in their segment like *Airbnb* in accommodation or *Uber* in passenger transportation. The study [6] estimates, that overall sharing revenue will potentially increase from 15 billion in 2015 to 335 billion by 2025. The drawback of sharing platforms in commercial use are the fees of up to 30% of the service price. Because of the high inefficiency due to empty trips, potential large quantities and relatively short detours in traditional long-distance logistics there is a margin for platform prices. Last mile instead is characterized by low quantities, limited cargo space, no dedicated return trips, short distances and complex tours with tight time constraints. These aspects result in much more complex matchmaking but lower gain, which additionally inherits the problem of micro-payments, in a scenario with even now precarious costs. Therefore, we propose an alternative by extending the platform functionality from a central operator to a distributed approach, controlled directly by the service provider and customers.

B. Blockchains in Logistics Applications

One method to distribute applications is distributed ledger technology like blockchains [7]. A blockchain is a linked list of data records that is continuously expanded. Cryptographic procedures are used to ensure that the concatenation and content of the blocks are permanently and immutably fixed. A certain consensus procedure allows several parties to establish and use a uniform database via a peer-to-peer (P2P) network without a proprietary central operator by means of blockchains. In addition to independence and the associated cost savings, blockchains offer further advantages. The distribution of the data can bypass technological bottlenecks of a central network structure and lead to higher reliability. Even if the blockchain itself is publicly accessible, suitable signature procedures can be used to ensure that only certain partners can access data. Due to the missing central organization of the partners, blockchains can be used anonymously by identification with a public key.

Since the publication of blockchains in [7] as part of a technology to implement the cryptocurrency Bitcoin, this technology was applied in various areas. A specific application, which has proved to be very promising is logistics as shown in the study [8] by Hackius and Petersen. Although here, blockchains are intended at a high level to ensure information management between various partners in long and complex supply chains. The benefits of the application are mainly seen in the ease of paperwork processing by using a consistently accessible data structure, identifying counterfeit products by verification or operating internet of things devices. In this paper however, we intend to use blockchains in a much more specific and operative scenario. We will show, that this technology can be the backbone of a P2P sharing network to overcome the problems of last mile logistics sharing mentioned in Section II-A.

C. Smart Contracts and Blockchain

In many cases, such as logistics, it has been shown that the concept of atomic transactions in the original blockchain implementation as introduced in [7] are not sufficient. Often business operations are based on prolonged interactions, which are controlled in complex processes and structures. For this purpose, some blockchain implementations like Ethereum [9] were extended by so-called smart contracts. Smart contracts are Turing-complete programs, which can be instantiated and used by blockchain users. The dependencies and characteristics of an application are modeled into methods, which operate on the blockchain and interact with network peers. The contracts are translated into a bytecode language and executed by an Ethereum Virtual Machine (EVM) on all nodes of the network. In this way, the integrity of the database is permanently ensured, as in the original blockchain approach. The execution of smart contracts costs a cryptoamount, which is clearly defined for each function and loosely coupled to the cryptocurrency ETH of underlying the Ethereum blockchain. This ensures efficient modeling of smart contracts and provides an incentive system for evaluation and block propagation. The in Ethereum common programming language Solidity allows a multitude of complex distributed applications (Dapps), such as independent organizational structures for NGOs, infrastructure for independent voting systems or generic platform economies [10]. In [11], Bogner et al describe an implementation of a sharing platform for rental services for example.

III. Distributed Multi-Use in Last Mile Logistics

The goal of this work is the implementation of a platform functionality in logistics via a combination of an adapted blockchain and smart contract technology as advancement to a sharing platform as illustrated in Figure 1. The focus is on the avoidance of a proprietary provider with corresponding costs due to infrastructure, service provisions and eventually the danger of reliance on a market monopoly position. Besides, there should be full control over data publishing, storage and a transparent process model through open smart contracts. Compared to a platform solution, the P2P approach should avoid the risk of infrastructure failure with increased performance and scalability.

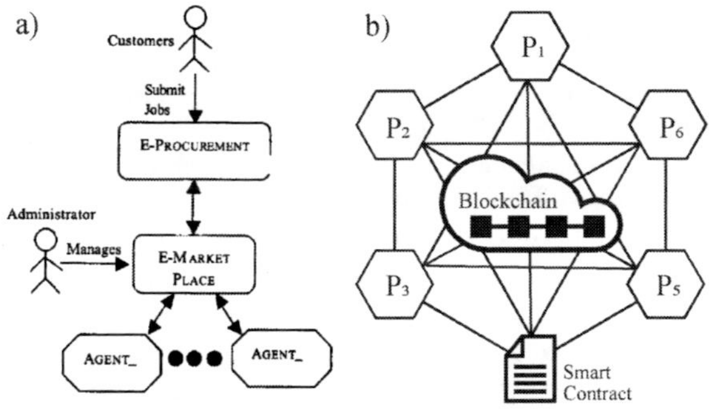

Figure 1. a) Architecture of a 4PL Sharing Platform from [5] compared to b) the distributed ledger P2P Network presented in this paper

A. Request and Offer Representation

The functionality of the network concentrates mainly on the intermediation of requests and orders between various partners. In this case, it covers all fundamental aspects of cargo sharing in a multi-use scenario. The roles of the partners can be distributed across all PL levels, as well as professional companies or private contractors. As usual in a last mile scenario, an order A consists of finite suborders $a_i \in A$. All a_i contain a tuple $< s_i, e_i >$ with the required departure and endpoint. Depending on whether A is a delivery or collecting tour, all departure or endpoints are usually identical and can be determined with s_A or e_A and the set of endpoints or departure points with E_A or S_A. A corresponding example scenario of three logistic partners and a request is illustrated in Figure 2.

Without loss of generality, we focus on delivery tours with a specific departure point s_A and set of endpoints E_A. A logistic partner $L_i \in L$ creates offers by integrating s_A and various subsets of E_A into their existing tours (or creating new ones) and value the costs with a function $o_i(\mathfrak{P}(E_A)) \mapsto \mathbb{R}$ over the power set of endpoint combinations. The matchmaking now finds a minimum set $\bigcup_i P_i = E$ of disjunctive subsets of E, so the sum of the cost of this partition $\sum_i o_i(E_i)$ becomes minimal.

Compared to real-world logistics requirements like weight, size, or temporal constraints, this basic request and offer model is not sufficient of course. However, it is a start for implementing a data structure in the smart contract for representing the fundamentals of interaction with and between blockchain peers.

B. Smart Contract Implementation

The corresponding smart contract and Ethereum client for a sharing functionality based on the model described above was implemented in Solidity 0.4.21 in a test environment provided by the Truffle Framework[12]. Truffle is a comprehensive toolbox for developing, testing and deploying smart contracts on a local Ethereum blockchain.

In the following, a minimal example will be explained to discuss various modeling decisions and correlations. First, a data model is declared in order to store and manage requests and offers on the blockchain via the contract. In addition to a generic data field, the status of a request, a mapping and an iterable array for offers are defined. The structure `Request`

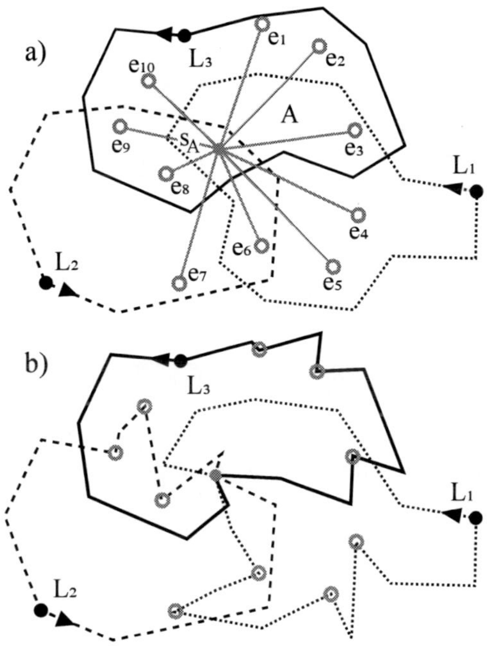

Figure 2. illustrated example of a) a request A in a logistics scenario and b) solution for tour integration

```
struct Request {
     string data;
     bool closed;
     bool completed;
     AcceptedOffers[] accepted;
     mapping(address=>string[]) offers;
     address[] offerers; }
```

Figure 3. Definition of a Data Structure for Requests

as defined in Figure 3 can now be used in the `placeRequest` method for initializing new requests as shown in Figure 4.

Placing a request involves the generation of a new `Request` instance, storing the data and calling a corresponding event linking to the request for notifying the P2P peers and potential contractors. The creation of offers works quite similar as shown in Figure 5. First, a `require` command checks the precondition of a valid open `Request`. Then, the new offer is attached to the `Request`.

The last example in Figure 6 illustrates the method for accepting offers. Preconditions of a valid `RequestID` for an

```
function placeRequest(string data) public
     returns (uint id) {
  uint RequestID = Requests[msg.sender].length++;
  Requests[msg.sender][orderID].data = data;
  emit NewRequestPlaced(msg.sender, RequestID);
  return RequestID; }
```

Figure 4. Method for Placing Requests

```
function placeOffer(address Requester,
    uint requestID, string data)
    public returns (uint id) {
  require(Requests[Requester].length > requestID);
  Request storage o = Requests[requester][requestID];
  require(o.closed == false);
  if(o.offers[msg.sender].length == 0) {
    o.offerers.push(msg.sender);
  }
  o.offers[msg.sender].push(data);
  return o.offers[msg.sender].length - 1; }
```

Figure 5. Method for Placing Offers

```
function acceptOffer(uint requestID,
    address offerer, uint offerID) public {
  require(Requests[msg.sender].length > requestID);
  Order storage o = orders[msg.sender][orderID];
  require(o.closed == true && o.completed == false);
  require(o.offers[offerer].length > offerID);
  uint index = o.accepted.length++;
  o.accepted[index].offerer = offerer;
  o.accepted[index].id = offerID; }
```

Figure 6. Method for Accepting Offers

open request as well as a valid `offerID` are checked. After that, the offer is added to the accepted offers of this request. These simple examples do not include a registration and signature check at the beginning to verify whether a user actually has access to the smart contract or a method. Only the originator is allowed to modify or withdraw requests or offers.

In the current state of the demonstrator, all data is stored in the blockchain for validation. However, it has been shown that this is not optimal in this application scenario. Due to a large number of possible offers, which depends on the number of elements in a request and the number of service providers, the storage effort increases exponentially. This leads to an inefficient execution of the smart contract, because storage operations on the blockchain are cost-intensive, and results into intense memory needs for storing the blockchain locally as well as higher network traffic for the propagation of new blocks. An implementation is currently being tested in which only requests are written to the blockchain and the offer phase is implemented via Whisper channels. The P2P users communicate directly with each other via the Ethereum specific communication protocol Whisper. Each client manages given and received offers independently locally. After the expiration of the request deadline the accepted offers calculated my the matching algorithm are stored on the blockchain and the contractors are notified via a corresponding event trigger or Whisper message. This contradicts the claim of a blockchain solution that all aspects of a contract are reproducible and verifiable stored but avoids the problem of storage effort and load.

C. The P2P Client Integration

In order to make the blockchain application as accessible as possible, an initial client implementation was created in connection with an Enterprise Resource Planning (ERP) system and user interface. A first approach primarily serves as a technological proof-of-concept. In the long term, the current status and history of the blockchain, as well as smart contracts, should be visualized user-friendly and operable. It also should be possible to define event triggers, enable automated requests deployments, evaluations and submissions of offers.

Based on the given Ethereum client implementation, an interface to a logistics ERP system was created, which contains a number of existing tours, orders and fleet information. This extended client is multifunctional with regard to a requester, as well as a offerer. The basic architecture is shown in Figure 7. A set of orders is converted from ERP data into a request specific format by the RequestHandler. The RequestHandler then publishes and manages the request on the blockchain via the smart contract methods as described above. Depending on the implementation, incoming offers are buffered by the RequestHandler locally or written on the blockchain directly by the contractor and evaluated by the RequestHandler at the end of the expiration time. The RequestHandler selects a set of offers based on predefined criteria in the business rule model and closes the request using the smart contract. The now implemented rules only choose offers according to minimize costs. Eventually, the smart contract triggers an event that notifies all bidders of the status of their bids.

In addition, the client reacts to events from the blockchain, such as newly generated external requests by the OfferHandler. This handler evaluates a request by incorporating each combination of subsets into existing tours from the ERP system as described in Section III-A. If there are possible tour combinations, the handler generates offers by evaluating the tour changes. The costs are calculated depending on the increased effort caused by the tour change and profit intentions. The effort is calculated using a Total Cost of Ownership (TCO) model that was developed in SDL specifically for logistics scenarios and whose integration is described in [13]. The profit intentions are stored in the business rule base along with other factors, such as under which circumstances an offer is to be submitted. The OfferHandler also evaluates blockchain events in case an offer is accepted to integrate the changes into the ERP system.

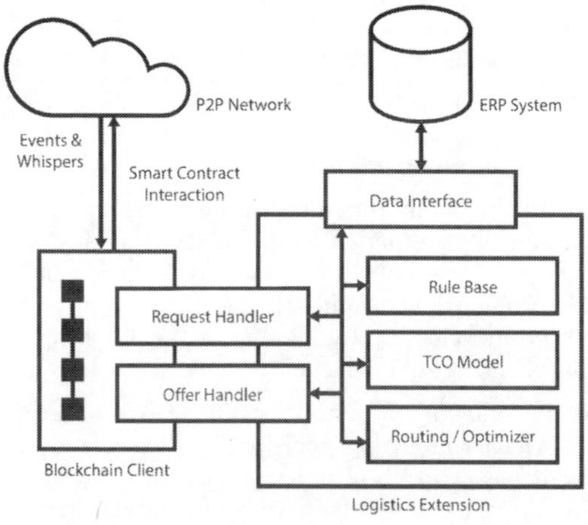

Figure 7. Components of the extended Blockchain Client

IV. Conclusion

In this work, we were able to present in a proof-of-concept that the marketplace functionality of a sharing platform for logistic goods can be implemented in the last mile context using blockchains. By doing so, service providers and clients are not dependent on a cost-intensive central proprietary platform but can conduct business via an autonomous consensual P2P network. The costs are therefore limited exclusively to the computing effort involved in expanding and storing the blockchain. The P2P approach ensures an easy ad hoc access for sharing resources and the distributed application implies high reliability and scalability. Eventually, the work on the demonstrator posed a few interesting modeling decisions and issues in multiple fields, which will be tackled in the future work within the SDL project.

V. Future Work

In addition to the development of a basic demonstrator, this work laid the foundation for further issues in research and development. Three basic problem areas were identified, which have to be covered for a successful application of a blockchain-based logistics sharing platform:

First, various aspects of the application domain must be clarified. The current model and methods are based on very simple and rudimentary assumptions. For a tangible application, a general uniform data model must be developed that covers all possible aspects of a last mile delivery, such as time, size or weight restrictions or additional requirements, such as cooling or tracking of transport. There is also the question of heuristics about the submission of offers, calculation of the profit share and which offers are accepted. Is a requester really looking for the cheapest solution or the smallest possible number of service providers to simplify scheduling and organization at the ramp? These issues are not blockchain specific but affect all sharing platforms in this application field. So it is foreseeable that under the research topic of 4PL there has to be an open standard covering data exchange aspects.

The question of legal regulations must also be addressed. The question of whether and how contracts can be (semi-)automatically concluded and under which conditions they are binding is currently still a research topic and far off from everyday use.

Eventually, there is the question of how and where the blockchain and thus the smart contract is deployed. A private blockchain with restricted access is conceivable and allows full control over deployment parameters and functionality but implies an organizational structure for registration and authentication. The implementation of such a central infrastructure would be opposed to the distributed concept of blockchains. An openly accessible but application-specific blockchain, on the other hand, requires a critical number of participants due to the danger of consensus attacks, which can compromise a blockchain. The third and most promising possibility is the deployment into the existing Ethereum blockchain. The main advantage here is the opportunity of using the established cryptocurrency ETH. This currency can be used directly in the logistics application to reimburse services or enable defining penalties in the contract. A disadvantage, however, is that there would be no control of aspects like the consensus algorithm for example and a heavy dependency on the actual ETH currency rate.

Acknowledgment

The project Smart Distribution Logistik (SDL), on which this paper is based, was funded by the Federal Ministry of Economics and Energy under the promotional reference number 01ME17001C. The responsibility for the content of this publication lies with the author.

References

[1] "Smart distribution logistik website," http://www.sdl-projekt.de, accessed: 2019-05-10.

[2] A. Juan, C. Mndez, J. Faulin, J. Armas, and S. Grasman, "Electric vehicles in logistics and transportation: A survey on emerging environmental, strategic, and operational challenges," Energies, vol. 9, 01 2016, p. 86.

[3] K. Stanoevska-Slabeva, V. Lenz-Kesekamp, and V. Suter, "Platforms and the sharing economy: An analysis eu h2020 research project ps2share: Participation, privacy, and power in the sharing economy, 2017," SSRN Electronic Journal, Nov 2017.

[4] L. Saglietto, "Towards a classification of fourth party logistics (4pl)," in Universal Journal of Industrial and Business Management, vol. 1, Jan 2013, pp. 104 – 116.

[5] H. C. Lau and Y. G. Goh, "An intelligent brokering system to support multi-agent web-based 4/sup th/-party logistics," in 14th IEEE International Conference on Tools with Artificial Intelligence, 2002. (ICTAI 2002). Proceedings., Nov 2002, pp. 154–161.

[6] PwC, "The sharing economy–sizing the revenue opportunity," 2015.

[7] S. Nakamoto, "Bitcoin: A peer-to-peer electronic cash system," 2009, accessed: 2019-05-10. [Online]. Available: http://www.bitcoin.org/bitcoin.pdf

[8] N. Hackius and M. Petersen, "Blockchain in logistics and supply chain : trick or treat?" in Proceedings of the Hamburg International Conference of Logistics (HICL), 2017, pp. 3–18.

[9] G. Wood, "Ethereum: A secure decentralised generalised transaction ledger," Ethereum project yellow paper, vol. 151, 2014, pp. 1–32, accessed: 2019-05-10.

[10] M. Mainelli and M. Smith, "Sharing ledgers for sharing economies: an exploration of mutual distributed ledgers (aka blockchain technology)," Journal of Financial Perspectives, vol. 3, no. 3, 2015.

[11] A. Bogner, M. Chanson, and A. Meeuw, "A decentralised sharing app running a smart contract on the ethereum blockchain," 11 2016, pp. 177–178.

[12] "Truffle suite framework website," https://truffleframework.com/, accessed: 2019-05-10.

[13] J. Kretzschmar, M. Johlke, and W. Rossak, "A tco analysis tool based on constraint systems for city logistics," in 7th International Conference on Advances in Vehicular Systems, Technologies and Applications, 2018 (VEHICULAR 2018). Proceedings., 2018, pp. 37–38.

Valuating and Pricing of Vehicle Generated Data as a Marketable Product in the Automotive Industry

Frank Bodendorf, Tobias Meissner and Jörg Franke
Friedrich-Alexander-University of Erlangen-Nuremberg (FAU)
Institute for Factory Automation and Production Systems
Erlangen, Germany
e-mail: frank.bodendorf@faps.fau.de
e-mail: tobias.d.meissner@fau.de
e-mail: joerg.franke@faps.fau.de

Abstract— **Data-driven business models play a significant role in the digital transformation of traditional value-added industries. More and more existing and potential partners of automobile manufacturers show interest in the vehicle data generated by their products. However, there is still no monetary value assessment to support decisions regarding the release of data. Traditional pricing approaches for material goods are based on cost, margin, and volume. However, these bottom-up calculation approaches are not applicable to digital goods. The background to this is, among other things, the uncertainty about the potential sales volume, the difficulty of cost splitting, and the high cost degression per unit of digital goods. This paper provides decision support for selling data to third parties as an intangible product. It introduces a concept that allows to value data generated by a motor vehicle in order to determine potential prospects and prices for sale. The evaluation model developed supports the car manufacturer's negotiating position towards potential data buyers.**

Keywords - Automotive Industry; Car Data; Business Model; Value Estimation

I. Introduction

The use of valuable data will fundamentally change competition in the future [21]. "The expected growth of the value pool from car data and shared mobility could add up to more than USD 1.5 trillion by 2030" [1]. Volume and quality of this "data treasure" will create strategic as well as operational competitive advantages [13].

Today, data is generated in large quantities by the vehicle, recording values of thousands of attributes. On the one hand, the vehicle user (driver) has the opportunity to enter data in on-board systems and "exchange" them for services. He/she receives individually adapted functions, such as voice control, comfort settings when entering the car, navigational instructions in real time or other services [3][4]. On the other hand, a variety of sensors and computers in the vehicle, unnoticed by the driver, generates a steady stream of data, which among others serves for control purposes [8]. Examples are the anti-lock braking system or the automatic windscreen wiper and light regulation. According to an internal study the data usage can be divided into nine purpose oriented categories:

- Facilitating vehicle use
- Meeting regulatory and legal requirements
- Supporting marketing and advertisement
- Assessing IT security
- Improving technical processes
- Fulfilling terms of contract
- Innovating and developing products
- Ensuring road safety
- Offering services to third parties

All of these categories have in common that value is created through the use of collected vehicle data. On the one hand, this value is reflected in technical or qualitative improvements as well as in cost reductions of the company's internal processes. On the other hand, the use of vehicle data can also lead to an economic improvement of the business results and in particular to an increase in turnover [22]. This may be a result of higher sales figures of products, i.e., manufactured vehicles, which are more attractive through data-based functions ("data infused products"). In addition, it is possible to offer certain data for sale as an end product itself [12].

This paper focuses on selling data to third parties. The demand for vehicle generated data depends on the benefit seen or expected by the buyer. From the perspective of the data provider, it is important to determine the value of the data in order to estimate the demand potential on an external market and to create appropriate pricing models.

The paper consists of three main parts. First, it gives an insight into the state of the art of data evaluation and points out weaknesses. Second, a new methodology is introduced that combines business model analysis and value estimation approaches. Third, a use case is presented which illustrates the model application and leads through the main process steps of value estimation.

II. Existing Valuation Approaches

The generic value of a data product in sales situations cannot be determined by a benefit that has already been realized, since the data is not yet being used by the buyer at the time of the transfer. Therefore, an evaluation must be based on probable and potential benefits [12]. This value can be estimated by using qualitative and quantitative methodological approaches. For a corresponding systematic value determination of vehicle data, a number of existing evaluation methods is outlined. This is initially done by a tabular overview in Section A, followed by a more detailed description in Section B and C and by a discussion of limitations and transferability in Section III.A. Literature often speaks of data and information without exactly differentiate between these terms. Some authors see in information "refined data", e. g., by placing it in a context of meaning. In this paper, both terms are used synonymously.

A. *Overview of potential methods for data evaluation*

The identification and selection of potential valuation procedures is done through a combination of literature review and in-depth interviews. First, 20 sources of literature are used to collect a comprehensive set of possible valuation approaches. Subsequently, valuation approaches are selected and specified with the help of 50 in-depth interviews with experts from the divisions or departments in the areas of cost engineering, data strategy, data analysis, and purchasing. The consolidated results are shown in Table I [18].

TABLE I. QUALITATIVE AND QUANTITATIVE VALUATION APPROACHES FOR DATA

Method	**Characteristics**			
	Type	*Input*	*Operator*	*Output*
Data Product Scorecard	Qualitative	Data attributes	Scoring-method	Willingness to pay
Data Value Design Canvas	Qualitative	Data use case	Expert workshop, Canvas nine factors	Interactions / connections
Value determi-nation per user	Quantitative	Acquisition cost Number of users	Discount calculation	Price per user-dataset
Value improve-ment by data services	Quantitative	Data material	Statistical analysis, e. g, hypothesis testing	Increase in value or quality through the use of data
Value determi-nation by Laney	Qualitative and Quantitative	Data material	Gartner Valuation Model	Qualitiative and financial value
Value determi-nation by partners	Quantitative	Theoretical value, maturity, expiration of information	Intangible Assets Evaluation	Monetary information value
Pricing based on customer value	Quantitative	Different data bundels	Versioning, price differen-tiation, surcharge calculation	Price for data bundels

B. *Qualitative evaluation*

The appraisal of methods has been carried out by literature review and interviews as mentioned in section II.A. The following most popular methods for evaluating vehicle data are identified:

- Data Product Scorecard
- Data Value Design Canvas

The *Data Product Scorecard* is a method of pricing on data marketplaces. For this purpose, the customer's willingness to pay depending on various data properties must first be estimated. This qualitative evaluation of the data properties is made by the Data Product Scorecard from a simulated perspective of end users or potential buyers of the data [20]. As part of an evaluation workshop within the company, the role of the user is taken and each data characteristic given in the scorecard is rated with 0, 5 or 10 points.

The *Data Value Design Canvas* approach looks at the data value chain. The approach is based on the theory of Service Dominant Logic and the "Jobs-To-Be-Done" theory [2][14]. According to [16], the data value chain begins with the generation of data and extends up to the provision of information to the (paying) customer.

C. *Quantitative evaluation*

Many companies have problems finding the real economic value of their data [15]. For a rethink in the development of new business models [7] and the optimization of internal processes, the determination of this value, especially for the automotive industry, is of particular importance.

Value determination per user

When acquiring companies with data-driven business models who have not yet monetized their database but still offer data-based applications to the end user, the data value is often determined by the value of the application per user. The price of acquisition is divided by the total number of end users of the application. From this calculated price per user, the average user acquisition costs are subtracted [10][19].

Value determination by Laney

According to Laney the data is evaluated through quality-based and quantitative financial analysis [12]. In the quality-oriented evaluation, the output is a scoring value between zero and one, in the financial evaluation an absolute monetary value. The two-part consideration focuses on methods for improving the "Information Management Discipline" and deals with "Foundational Measures" as

- How correct, complete and exclusive is the data? (Intrinsic Value),
- How good and relevant is the data for specific purposes? (Business Value),
- How does this data affect key business drivers? (Performance Value).

On the other hand, the "Information Economic Benefit" of "Financial Measures" is examined:

- What would it cost us if we lose this data? (Cost Value),

- What could we get from selling or trading this data? (Market Value),
- How does this data contribute to our bottom line? (Economic Value).

Both considerations provide a quantitatively measurable contribution to the value of data and will be explained in more detail below. Based on a collection and analysis of existing valuation approaches according to Laney, the combination of a plausibility check and a requirements analysis leads to a new valuation perspective introduced into a new model presented in chapter III.

III. Developed Methodology

A. *Motivation of a new evaluation approach*

Value Determination by Laney comprises pure comparative methods. These do not calculate monetary values, yet they show some interesting perspectives. The Data Value Design Canvas lists nine factors which affect the value of data. The effects of information or data are considered generally. For example, information/data protects against unwanted events or promotes wanted events. Unwanted events always result in costs. Thus, the avoidance of unwanted events corresponds to a cost reduction. The realization of desired events effects an increase in sales as the most relevant example. The Data Product Scorecard assesses the willingness to pay of a customer. If the information considered is "perfect", the customer is willing to pay the full price for this information.

The method according to Laney, the Data Value Design Canvas, and the Data Product Scorecard focus on the quality of the information or data. Both methods have in common that always certain use cases of data usage are considered.

By comparing the different approaches listed in Section II, some requirements can be derived for a new concept integrating different aspects:

- Quality factors must be taken into account.
- The willingness to pay is relevant for data sales. This depends on various factors, including the purchase motive, the perceived benefit, the reputation of the seller, and the individual purchase situation.
- Data has the potential to increase sales at the customer or to reduce costs for internal customer processes.
- Competition should be considered as an important factor.

B. *Evaluation model*

The developed "integrated methodology" for an innovative evaluation model meets these requirements from the fields of quality assessment [17], price differentiation [24], cost management, and competitive analysis in a combined way [22]. So, for a specific use case, it is possible to estimate a monetary value of data by integrating different aspects (see section III.A). Basically, Laney's approaches are not limited to any specific field of application [23].

The plausibility check of selling prices is achieved by combining qualitative tools based on methods such as Business Model Canvas or Data Canvas with a practical evaluation through quality workshops as well as quantitative calculations. These include, among other things, the valuation by Laney, a bottom-up cost calculation as well as profit split approaches.

The process model outlined in Figure 1 shows the process steps of the model for a use-case-specific value determination. The non-rivalry property of data enables multiple sales of similar data bundles or even the same dataset. The total value of the data bundle can be determined as the sum of the values across all (potential) use cases:

$$V_G = \sum_{i=1}^{n} V_i \qquad \text{(a)}$$

V_G	Total value of data bundle G
V_i	Individual value of data for the use case i
i	Use case index
n	Number of use cases

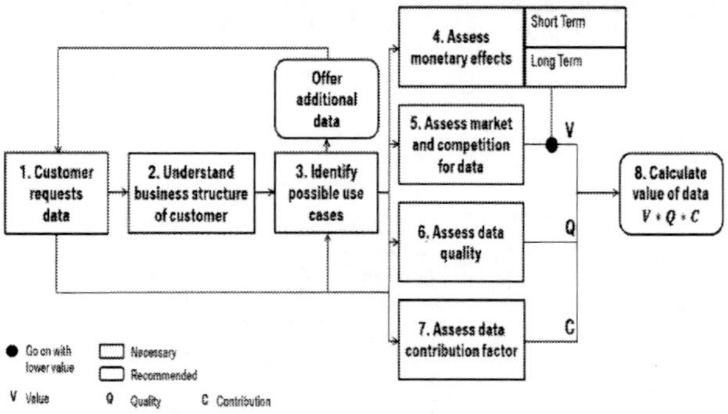

Figure 1. Process for the monetary valuation of data

As a first step in the process, data requests from potential business customers are collected or data is proactively offered to the customer. In order to be able to identify potential data needs, the customer's business model must first be understood. (1 + 2)

The way the customer translates the data into benefits can be identified through a systematic analysis of possible use cases. For this a combination of the Business Model Canvas and Data Value Design Canvas is suitable. With progressive understanding of the application, it is theoretically possible to offer different or additional versions of data bundles to the customer. (2 + 3)

In order to determine the customer's willingness to pay, the value of the use case must be understood in detail [9]. For this, a possible cost reduction or increase in sales by the data is to be determined.

For each use case, there is both a short-term and a long-term monetary benefit, which in individual cases can also be zero. The model of Figure 1 shows a parallel approach to Laney's business value calculation of information, which determines the data relevance for specific processes. (4)

In cases where there is competition on "data marketplaces" or the self-collection of data is significantly more favorable than granting the monetary benefit to a third party, these influences must be measured for further calculation. (5)

At the same time, the data is qualitatively evaluated based on selected criteria (see section II.B). The model considers the monetary value of quality criteria, following Laney's valuation ideas in the form of a quality factor Q. (6)

In addition, the data contribution factor C takes into account that other vehicle generated data or additional information may be necessary in addition to the offered vehicle generated data. (7)

After having carried out steps 1 to 7, finally the use-case-specific combined value of the data is calculated (8). The determined preliminary data value V is multiplied by the quality factor Q and the data contribution factor C. Both of them are discount factors. The calculations of Q, C, and V are illustrated in section IV. Hence, V is reduced according to low quality or insufficient amount of data. Interdependencies between the factors are possible but not taken into account. E. g., low data quality can lead to a higher demand for data from other sources. A lowered data quality factor can therefore further reduce the data contribution factor.

IV. USE CASE

The applicability of the developed methodology is experimentally tested for real sales situations. One of them is the offering of Road Segment Data (RSD) to a navigation maps provider. This data is needed to provide a high definition roadmap for autonomous driving.

A. Business structure analysis

The results of the structural analysis of the business case are outlined below. On this basis, further analysis leads to the data needs and the data bundles to be offered [11][20].

- *Value proposition:* build tomorrow's road network
- *Customers:* companies
- *Segments:* automotive industries
- *Channels:* direct contact
- *Revenue stream:* Selling refined information (HD map) for autonomous driving and location based services
- *Cost structure:* personnel, data transfer
- *Key resources:* street/road data, navigational data, real time traffic information data, road segment data (RSD)
- ***Data offering:*** road segment data captured by car cameras
  (edge markings, center markings, strip width, crash barriers, guide posts, signs, wild animal warning reflectors and barriers)

B. Monetary Effect

An internal study of the car the manufacturer says that the willingness to pay for the enriched HD map in the self-driving vehicle industry is 60 € per year for a highly autonomous or fully autonomous vehicle. For the year 2020, a global volume of 3.3 million high or fully autonomous vehicles is expected on the market. In 2035 this number is predicted to be 28 million vehicles. The market for navigation maps in vehicles is divided into market shares of 15% to 25% [6].

These assumptions lead in the worst case scenario (assuming a market share of 15%) to a potential turnover of at least 29.7 million € in 2020 (3.3 million * 60 € * 0.15) and 252 million € in 2035 (28 million * 60 € * 0.15).

C. Market competition

The market position is qualitatively described by the criteria "is it valuable", "is it rare", "is it hard to imitate", and "is the firm organized for success" according to [15]. The answers to the questions in a conducted survey are based on a competitive analysis of 26 navigation maps suppliers. The qualitative assessment shows that there is a "short term competitive advantage".

D. Data quality and contribution factors

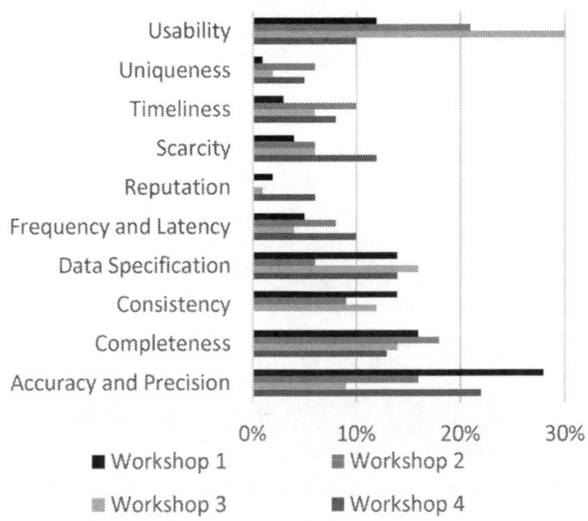

Figure 2 Evaluation of the quality criteria of data

In order to evaluate the data quality, 14 data scientists are interviewed in four expert workshops on given quality criteria. The results of the workshops are summarized in Figure 2 and transformed into a quality factor Q.

$$Q = \sum_{i=1}^{n} c_i * w_i \qquad \text{(b)}$$

Q	Quality factor
c_i	Evaluation factor of criterion i
w_i	Weight of criterion i
i	Index of criterion
n	Number of criteria

The evaluation factor values c_i of the criteria result from a pairwise comparison of all criteria in a preference matrix. The data contribution factor C expresses if all required data (contribution factor = 1), almost all data (contribution factor = 0.75), about half of the data (contribution factor = 0.5), few data (contribution factor = 0.25) or no data (contribution factor = 0) can be provided.

$$C = CM * w_{cm} + CK * w_{ck} \qquad \text{(c)}$$

C	Data contributing factor
CM	Contribution factor of metadata
CK	Contribution factor of key data
w_{cm}	Weight of metadata
w_{ck}	Weight of key data

The use-case-specific factors CM und CK differentiate between key data und metadata (additional data). The expert based weights w_{cm} and w_{ck} rate the relative data contribution of each data type in the use case. Location based services are one example. They are based on customer preferences (metadata) on the one hand and GPS data (key data) generated by the vehicle on the other hand. Experts give weights of 0.2 for GPS data and 0.8 for preference data. It is assumed that the vehicle can deliver 90 percent of the GPS data needed, however none of the preference data. So, the contribution factor of the metadata is zero. The calculation of C results in the value of 0.18:

$$0.2 * 0.9 + 0.8 * 0 = 0.18$$

The corresponding calculation for the navigation maps provider gives the following result:

Quality factor Q = 0.8

Contribution factor C = 0.55

E. Selling prices

Through the product of monetary effect (*V*), quality (*Q*) and contribution factor (*C*) (see Figure 1) the value for the offered data for the use case "navigation map provider" is in the worst case scenario 13.07 million € (29.7 million € * 0.8 * 0.55).

F. Lessons Learned

On the one hand the presented evaluation model has been applied to several fictitious use cases with real information coming from companies interested in buying data but without any sales decision. On the other hand the evaluation model was tested on several specific sales situations and the outcome of the model, i. e. the post calculated data value, was compared to the real sales price. The monetary data values determined using the evaluation model show an average deviation of 8% from the sales prices negotiated in practice.

Thus, the created model seems to deliver a good approximation to price imaginations which are based on gut instincts so far. The model is capable to support price negotiations by providing a systematic methodological basis and a transparent multidimensional valuation framework.

V. CONCLUSIONS

The methods of data evaluation identified in the literature are individually not suitable for practical value determination of data and their pricing in sales situations. This paper presents a methodology that focuses on the selling of data as intangible products to external business partners.

The methodology can also be transferred to use cases within the company. In addition to determining the value of the data, decisions regarding the pricing model must be made.

However, for long-term strategies it is unclear to what extent currently recorded data is valuable in the future. Data that is still useless, because currently there are no use cases, can be highly relevant for future use cases. Because of the existing knowledge gap and missing empirical values, it is impossible to determine a value of data over the entire lifecycle, above all because of very uncertain future potentials. This article exemplifies a possible evaluation and monetization of a small fraction of the total data available in the automotive industry.

Against the background of the huge amounts of data available there, it quickly becomes clear that due to technical limitations probably never all potential use cases can be implemented. There are various transmission options for vehicle generated data. The built-in memory can be read in authorized garages, updates can be transmitted at weekly or daily intervals, or data can be transferred in real time. There is always a technical limitation due to the restricted transfer rates or transfer options. Not all conceivable applications can be realized at the same time.

It is also an open question whether it makes sense to regard the vehicle as an open platform. In this case, an automobile manufacturer or even the automotive industry as a platform provider could probably sell the platform as a service (PaaS) to service providers who will pay for specific data accessed via the platform. As an analogy, platforms of Apple and Android can be considered. Third parties develop services to be offered on these platforms. The developed services (e. g., apps) increase the attractiveness of the platform. Depending on the design, there are direct and indirect network effects. With regard to autonomous driving, this approach may potentially increase the attractiveness of vehicles and vehicle fleets acting as such platforms. For example, in addition to many existing connected drive services, applications of third-party providers can be activated, which leads to an immense increase of the value of a ride and the driving experience for the customer. Here, completely new service ecosystems spanning and connecting different industrial sectors are appearing. To name only one step toward the future, the intelligent personal assistant from BOSCH enables the networking of car services and e-home services [5].

REFERENCES

[1] J. Balasubramanian, S. Beiker, S. Chauhan, T. Colombo, F. Hansson, S. Inampudi, R. Jaarsma and M. Kässer, "Car data: paving the way to value creating mobility : Perspectives on a new automotive business model", Advanced Industries McKinsey, 2016

[2] L. A. Bettencourt and A. W. Ulwick, "The customer-centered innovation map", Harvard Business Review, vol. 86 (5), pp. 109-114, 2008

[3] BMW: *BMW ConnectedDrive Kundenportal – digitale Vernetzung zu Ihrem BMW.* [Online]. Available from: https://www.bmw-connecteddrive.de/app/index.html#/portal/store/Base_TolCarOffer [retrieved: 11, 2018]

[4] BMW: *NOW Mobilitätsdienstleistungen.* [Online]. Available from: https://www.bmwgroup.com/de/marken/now-mobilitaetsdienstleistungen.html[retrieved: 01, 2019]

[5] BOSCH: *Nur Fahren war gestern – der persönliche Assistent ist morgen.* [Online]. Available from: https://www.bosch-presse.de/pressportal/de/de/nur-fahren-war-gestern-%E2%80%93-der-persoenliche-assistent-ist-morgen-101568.html. [retrieved: 01, 2019]

[6] A. Brugger: *Global production forecast for semi- and fully automated vehicles.* [Online]. Available from: https://www.oliverwyman.com/con-tent/dam/oliver-wyman/global/en/files/who-we-are/press-releases/Oli-verWyman_Graphics_Value%20Pools%20Autonomous%20 Driving_EN_16072015_final.pdf [retrieved: 11, 2018]

[7] J. L. L. Francisco and J. Esteves, Value in a Digital World: How to assess business models and measure value in a digital world, Berlin: Springer, 2017. IS BN 978-3-319-51750-6

[8] W. Gründinger: *Data-Driven Business Models in Connected Cars, Mobility Services & Beyond.* [Online]. Available from: https://www.bvdw.org/fileadmin/bvdw/upload/publikationen/connected_mobility/20180418_data_driven_business_models_Seiberth_Gruendinger.pdf [retrieved: 01, 2019]

[9] R. Harmon, H. Demirkan, B. Hefley , and N. Auseklis, "Pricing Strategies for Information Technology Services: A Value-Based Approach", 42nd annual Hawai'i International Conference on System Sciences (HICSS 2009), Piscataway, NJ: IEEE Press, Jan. 2009, pp. 1–10, ISBN: 978-1-424-44197-6

[10] N. Henke, et al., The Age Of Analytics: Competing In A Data-Driven World, McKinsey Global Institute, 2016

[11] HERE: *Location-Based Services for the Autonomous Future.* [Online]. Available from: https://www.here.com/en. [retrieved: 01, 2019]

[12] D. B. Laney, Infonomics, How to monetize, manage, and measure information as an asset for competitive advantage, New York, NY: bibliomotion inc, 2018, ISBN: 978-1138-09038-5

[13] E. Leonidas, DataStreamX: *A Practical Guide to Pricing Data Products.* [Online]. Available from: http://cdn2.hubspot.net/hubfs/573334/Downloadable_Content_(WP_or_Guides)/DataStreamX_Data_Product_Pricing_Whitepaper.pdf [retrieved: 11, 2018]

[14] C. Lim, et al., "From data to value: A nine-factor framework for data-based value creation in infor-mation-intensive services", International Journal of Information Management, vol. 39, pp. 121–135, 2018, ISSN: 0268-4012

[15] S.M. Liozu, W. Ulaga, Monetizing Data, A Practical Roadmap for Framing, Pricing & Selling Your B2B Digital Offers, Guetersloh, Germany: VIA Publishing, 2018, ISBN: 978-1-945815-04-1

[16] K. Mathis and F. Köhler, "Data-Need Fit - Towards data-driven business model innovation", The Fifth Service Design and Innovation conference (ServDes.), pp. 458–467, 2016, ISBN: 978-91-7685-738-0

[17] D. Mcglivary, "Executing Data Quality Projects: Ten Steps to Quality Data and Trusted Information: Definitions of Data Categories", Chief Data Officer & Information Quality Symposium, 2009

[18] T. Meißner, Wert und Verwertung von Daten der Hardware- und Softwarenutzung im Fahrzeug, University of Erlangen-Nuremberg, Master Thesis, Erlangen, 2018

[19] K. O'Neal, Quantifying the Value of Data: The First Step in Data, unpublished presentation

[20] A. Osterwalder and Y. Pigneur: Business Model Generation: A Handbook for Visionaries, Game Changers, and Challengers, Hoboken, New Jersey: John Wiley & Sons, 2010, ISBN: 978-0470-87641-1

[21] B. Schmarzo and M. Sidaoui, Applying Economic Concepts To Big Data To Determine The Financial Value Of The Organization's Data And Analytics, And Understanding The Ramifications On The Organizations' Financial State-ments And IT Operations And Business Strategies. [Online]. Available from: https://in-focus.dellemc.com/wp-content/uploads/2017/04/USF_The_Economics_of_Data_and_Analytics-Final3.pdf [retrieved: 01, 2019]

[22] C. Shapiro and H. R. Varian, Versioning, "The Smart Way to Sell Information", Harvard Business Review, vol. 6, pp. 107 f., 1998

[23] F. Stahl and G. Vossen, "High quality information provisioning and data pricing", 29th International Conference on Data Engineering Workshops (ICDEW 2013), 2013, pp. 290-293

[24] S. Viswanathan and G. Anandalingam, "Pricing strategies for information goods", Sadhana Indian Academic Sciences Journal, vol. 30, 2005, pp. 257–274, ISSN: 0256-2499

Wirtschaftswoche: *GPS-Satellitennavigation: Wie Navigationsgeräte-Firmen um einen Milliarden-Markt kämpfen.* [Online]. Available from: https://www.wiwo.de/unternehmen/gps-satellitennavigation-wie-navigationsgeraete-firmen-um-einen-milliarden-markt-ka-empfen-seite-3/5426542-3.html [retrieved: 11, 2018]

Real Deployment of the V2X-Based Data Probe Application and its Integration with a Commercial Traffic Platform

Hamid Menouar and Mohamed Ben Brahim
Qatar Mobility Innovations Center
Qatar University
Doha, Qatar
e-mail: hamidm@qmic.com, mohamedb@qmic.com

***Abstract*—In Vehicle to vehicle and vehicle to infrastructure communication the vehicles are expected to transmit continuously and frequently information that include their velocity and location coordinates. Such real-time geographical information, if collected properly, can be used for inferring the current status of the traffic on the roads. In this contribution, the authors elaborate on how this information can be used to enable an accurate source of data for traffic management platforms. The contribution includes a real implementation of the solution using a standard-compliant V2X platform, as well as its integration with a commercial traffic platform.**

Keywords-V2X; V2V; V2I; C2X; DSRC; VANET; probe; traffic; ITS.

I. INTRODUCTION

Vehicle to vehicle and to infrastructure communication (V2X) is an emerging technology that is expected to bring on our vehicles' dashboards many new interesting applications and services [1]. Certainly, road-safety applications are the key driver for such an emerging technology, but there are also many infotainment and traffic efficiency applications that will be enabled on top of such a technology.

Indeed, when V2X will be deployed, many new applications and improvements of existing applications become possible. Traffic information is a good example of existing applications that will be improved considerably thanks to V2X.

In V2X related standards [2], Data Probe is listed as one of the first applications that will be enabled by V2X. In such an application, selected vehicles participate by sharing their probe data with a designated Road Side Unit (RSU) deployed on the road. In this contribution we present our implementation of the Data Probe application along with its integration with a commercial traffic platform.

The rest of this paper is organized as follows: Section II presents the related works. Section III presents the implemented solution. Section IV presents the test-bed and field tests. Finally, Section V concludes the contribution.

II. RELATED WORK

A. Tradditional Traffic Data Calculation

Different techniques have been used in the field of Intelligent Transport Systems (ITS) to calculate close-to-real-time traffic information [3]. Such techniques rely mainly on collected data about road users that come from sensors installed either at fixed locations on the road, in road user's vehicles, or in road users' mobile devices [6]. These collected probe data is used to calculate the traffic information and enable different ITS services and applications. One of the simplest solutions to calculate the traffic information from probe data consists of comparing the reported road user's speed to the normal speed of the road segment on which the user is currently driving.

B. V2X-based Data Probe

Data Probe is one of the applications listed in the V2X-related standards [2], and it consists of using V2X capabilities to collect accurate and real-time data about the vehicles on the road. As described in [2][4], to enable the data probe application, the On-Board Units (OBUs), while driving on the road, report their current and past probe data (location, speed, heading, etc.) to a nearby RSU. After receiving the probe data from passing vehicles, the RSU can transmit it to the backend. Once the data reaches the backend, it is stored and processed to produce real-time and historical traffic information.

As per the related standards [2], the transmission of the probe data by the passing vehicles can happen in three ways: (1) periodically: transmitted every defined period of time, (2) reactively: transmitted as a response to a request received from the RSU, and (3) while moving: transmitted only when the vehicles is moving.

Only some vehicles participate in running the data probe application [2], which leads to some gaps due to the lack of information about other vehicles on the road. In [5], the authors propose to have the selected vehicles reporting not only theirs, but also their neighbors' probe data to the RSU through multi-hop communications.

III. USING BSM AND CAM TO ENRICH V2X-BASED DATA PROBE APPLICATION

In our solution we propose to use the Basic Safety Message (BSM) [2] or the Cooperative Awareness Message (CAM) [7] to collect the probe data of the passing vehicles. The BSM and CAM contain the location, speed, heading and other useful information, and they are transmitted periodically by each vehicle to inform other vehicles about its presence. Therefore, by using the information in the BSM and CAM received from neighboring vehicles, a selected vehicle can construct the probe data of all those neighboring vehicles. That selected vehicle can now include its neighbors' probe data when transmitting its own probe data to the RSU.

Our implementation is similar to the solution proposed in [5] with two main differences: (1) the reporting vehicles act in a reactive and not in a proactive mode, and they are assumed to be pre-equipped (e.g. belonging to a partner fleet), and (2) the transmission of the probe data is made through one and not multi hop.

By letting the vehicles that participate in the data probe application send theirs and their neighbors' probe data, the RSU can get a broader visibility about a larger number of vehicles in the traffic around its location.

Figure 1 illustrates the proposed solution through an example where A is a participating vehicle running the data probe application. The vehicle A constantly collects the probe data of other vehicles around it through their BSMs or CAMs. When entering the RSU's communication range, the vehicle A transmits to that RSU the collected probe data, which contain its own and its neighboring vehicles' probe data. The driving direction of the reported vehicles (heading field in CAM and BSM) is used in the backend to map each vehicle's probe data on the right road segment.

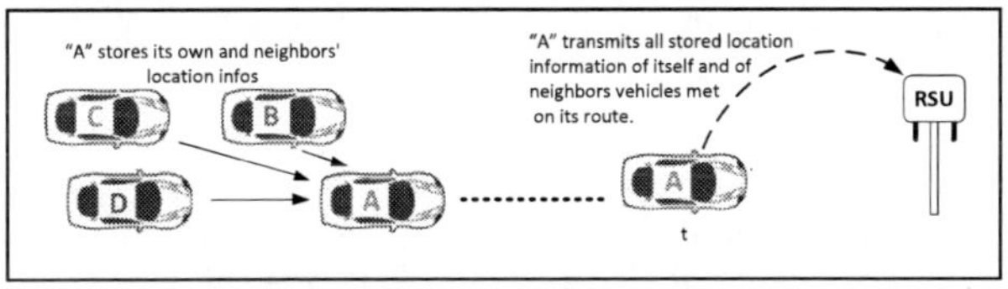

Figure 1. Example of a vehicle reporting the prob data of its own and of other vehicles in its communication range.

IV. IMPLEMENTATION AND FIELD EXPIRMENTS

To validate our contribution we have implemented and deployed it on a real V2X platform that is being piloted in Doha, Qatar. Our experimentations included three vehicles, each equipped with one OBU, and one RSU installed on the side of the road at a selected intersection. Both the OBUs and the RSU communicate over 5.9 GHz band in compliance to the European V2X standards [7]. In our implemented solution, a selected vehicle is equipped with a data probe application that keeps storing the probe data of its own and of any vehicle met on the road thanks to the received CAMs.

To validate our implementation from an end-to-end perspective, we interfaced it with our commercial traffic platform Masarak™ [8]. On Masarak, we can see both real-time and historical traffic information that is calculated based on the collected V2X probe data. Figure 2 illustrates snapshots of the obtained traffic information as shown by Masarak™ [8] on a map using road segment coloring codes (Red: heavily congested traffic, Orange: slow-moving traffic and Green: clear traffic).

In this initial experimentation work we did not pay attention to measuring the impact of the data probe application on the overall network overhead. This is something we plan to do along with other measurements in the future, and present the findings in an extended version of this contribution. But, we believe that the impact on the network overhead is relatively small, as only pre-selected vehicles are enabled with the data probe application. It is also important to mention that as per the standards [7], the traffic efficiency applications are assumed to use a different radio channel than the one allocated for safety applications.

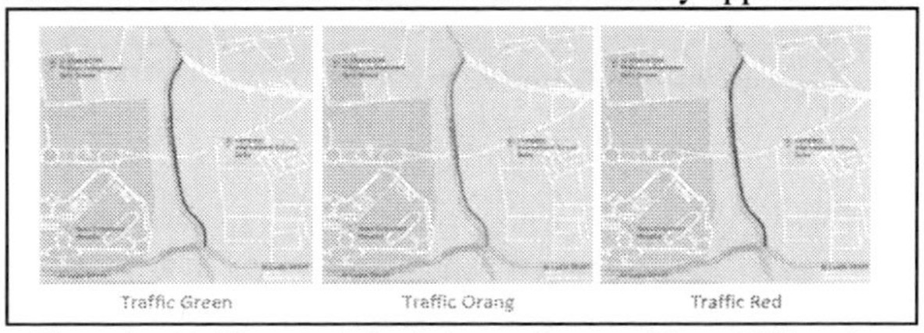

Figure 2. Traffic states information as shown on Masarak™ based on V2X probe data messages.

V. CONCLUSIONS AND FUTURE WORK

This contribution presents initial findings from our experience with the implementation of V2X-based Data Probe application and its integration with a commercial traffic platform. In our contribution, we proposed an enhancement to the conventional solution as described in the related standards, by letting a participating vehicle report not only its own probe data but also those of its non-participating neighboring vehicles through single-hop transmissions.

Further experiments and results of the implemented data probe application will be presented in a future longer paper.

ACKNOWLEDGMENT

This publication was made possible by NPRP grant #NPRP8-2459-1-482 from the Qatar National Research Fund (a member of Qatar Foundation). The statements made herein are solely the responsibility of the authors.

REFERENCES

[1] J. Harding, et. all., (2014, August) "Vehicle-to-vehicle communications: Readiness of V2V technology for application." (Report No. DOT HS 812 014). Washington, DC: National Highway Traffic Safety Administration.

[2] Dedicated Short Range Communications (DSRC) Message Set Dictionary, J2735_201603, 2016-03-30, issued by V2X Core Technical Committee, SAE International.

[3] L. Montero, et al., "Impact on Network Performance of Probe Vehicle Data Usage: An Experimental Design for Simulation Assessment," Journal of Advanced Transportation, vol. 2018, Article ID 3736417, 12 pages, 2018.

[4] M. B. Brahim and H. Menouar, "V2X-based traffic flow calculation with support of unique identifier randomization," 2016 2nd IEEE International Conference on Computer and Communications (ICCC), Chengdu, 2016, pp. 1226-1232.

[5] L. Yang, J. Xu, G. Wu, J. Guo, "Road probing: RSU assisted data collection in vehicular networks." 2009 5th International Conference on Wireless Communications, Networking and Mobile Computing. IEEE, 2009.

[6] M. R. Islam, N. Ibn Shahid, D. T. Karim, A. Al Mamun, M. K. Rhaman, "An efficient algorithm for detecting traffic congestion and a framework for smart traffic control system," 18th International Conference on Advanced Communication Technology (ICACT), Pyeongchang, 2016, pp. 802-807.

[7] Intelligent Transport Systems (ITS); Vehicular Communications; Basic Set of Applications; Part 2: Specification of Cooperative Awareness Basic Service, ETSI TS 102 637-2 V1.2.1 (2011-03).

[8] Masarak™ - http://www.masarak.com/ [retrieved: May, 2019]

Lane Estimation Algorithm Based on Sensor Fusion Database

Seung uk Jeon
School of Mechanical and Automotive Engineering
Kyungil University
Gyeongsan, South Korea
Email: jsukk94@naver.com

Byungyong You
School of Mechanical and Automotive Engineering
Kyungil University
Gyeongsan, South Korea
Email: zealot@kiu.kr

Abstract— **This paper presents a lane estimation algorithm to recognize the position of the vehicle and the total number of lanes on the road. For more efficient self-driving and ADAS, it is important to recognize ego-vehicle position using environmental sensors. This algorithm provides more convenient services based on precise position of ego – vehicle. The study was based on an actual driving database and algorithm is designed according to various road conditions. The test results with actual driving data showed high accuracy of proposed algorithm.**

Keywords- Recognition; Judgement; Algorithm; Lane Estimation; Accuracy;

Ⅰ. INTRODUCTION

Advanced Driver Assistance Systems (ADAS) have been commercialized to enhance convenience and safety for motorists driving cars, and the prevalence of streamlined electric systems worldwide is increasing. Location determination technology that estimates the location of cars is essential for self-driving cars and ADAS. The sensorconverged positioning system uses a method to estimate the location of the ego-vehicle by converging precision maps with the environment-aware sensors (camera, lidar, radar, etc.) along with Global Positioning System (GPS) and Inertial Measurement Unit (IMU).

Once the location of the ego-vehicle is known correctly, a variety of technologies, such as environmental awareness, vehicle control optimized for the surrounding terrain, and change of driving strategy according to the terrain can be carried out more effectively. One of the most widely used location determination technologies is the radio navigation system that uses satellite information, such as GPS. While this method provides an absolute position in the earth's coordinate system and has the advantage of not accumulating depends on the radio wave receiving situation. To compensate for the increasing number of errors over time, inertial navigation methods such as IMU are being used. Although this approach has the advantage of providing a precise relative position at a short distance, regardless of the radio reception situation, the problem is that the error continues to increase with time due to the limitations of the cumulative-based position estimation method. To overcome the limitations of these existing methods, a recent active study of sensor fusion based precision measurements is performed in [1] and [4]. In general, sensor fusion based precision vehicle location system uses a method to estimate the location of ego-vehicle by converging precision maps with environmental recognition sensors (camera, lidar, etc.) along with GPS and IMU. However, studies have not yet been done to estimate the lane of ego-vehicle in these papers. So, this paper focused on estimating the lane of ego-vehicle and carried out the study. Pre-recognition of the lane information that the ego-vehicle is driving will help the driver of the normal driving, or the driver of the self-driving control system to provide more efficient driving. To further enhance the performance of the lane estimation algorithm, a study was made in [3] on the direction of improvement of lane assessment that provided an integrated framework for lane estimation clues.

Existing studies [1]-[4] used a sensor fusion based precision vehicle location system to pinpoint the position of the ego-vehicle, but proposed algorithm designed a algorithm to estimate the position of the lane under which dynamic obstacles are located on the roadway's lane and the position of the carriage is unrecognizable. Algorithm designed in this study uses relative vehicle information from pre-collected database to estimate the lane of the egovehicle based on lateral distance from dynamic obstacles around the ego-vehicle. For the verification of algorithm, the performance evaluation was conducted on the actual driving environment data of eight roads extracted from the database, and it was found that the lanes of the ego-vehicle could be estimated relatively accurately even when the total number of lanes of the road is not known.

Ⅱ. DATABASE CONSTRUCTION

The vehicles used in this study are constructed from Figure 1. by attaching environmental sensors (LIDAR, RADAR, and CAMERA) to the KIA Carnival vehicles. The Spatial Information Research Institute collected actual driving data from the Ansan Expressway in South Korea to the set up the database and provided this database in order to support the algorithm.

The sensors that are mounted include environmental recognition sensors (five Lidar, three Radar, six Camera) and location recognition sensors (two INS/DGPS [VRS RTK], and two DMI and OBD). The information used primarily in this study is the x-axis lateral coordinates. Extracted the relative vehicle coordinate information of around the ego-vehicle shows that it has relative distance and direction from the ego-vehicle.

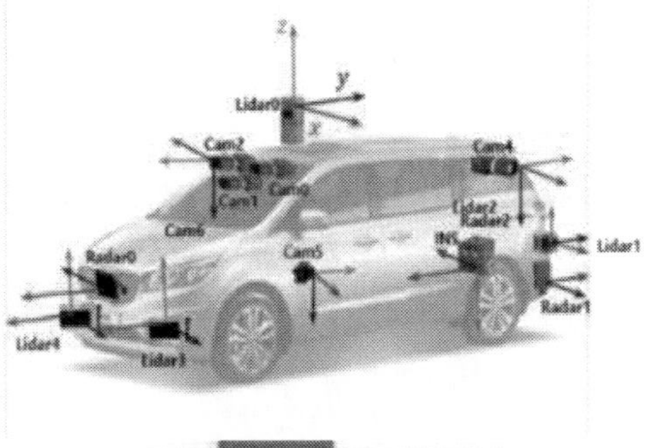

Object_ID	CenterX	CenterY	CenterZ
29	-10	-955	-149.582
41	50	1175	-73.2534
87	-385	-3280	-216.014
90	750	2460	-37.3392
91	20	1990	-4.68465
92	-380	-2495	-128.359
93	355	1660	-69.7688
94	-310	-5120	-264.192
95	695	1015	-93.9234

Figure 1. Data Acquisition System & X coordinates information

Ⅲ. ALGORITHM DEVELOPMENT

The components of the coordinate system set in the reference car are directional with **x**, **y** and **z**, and these coordinate systems are used to recognize the dynamic obstacles around the ego-vehicle. Algorithm uses the lateral components of the x-axis coordinates and the relative vehicle information recognized around the ego-vehicle.

The difference in distance between the self-vehicle and the perceived center of the adjacent relative vehicle is shown in Figure 2. A condition-based algorithm was developed to estimate the lane where the ego-vehicle is located by extracting the distance difference between the ego-vehicle and the recognized adjacent relative vehicle.

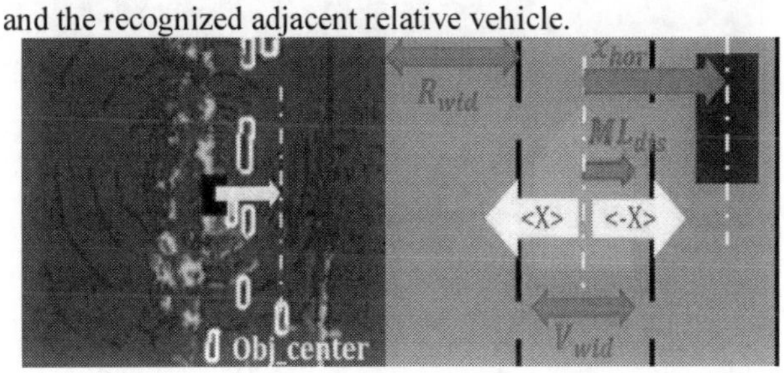

Figure 2. Algoribthm development environment

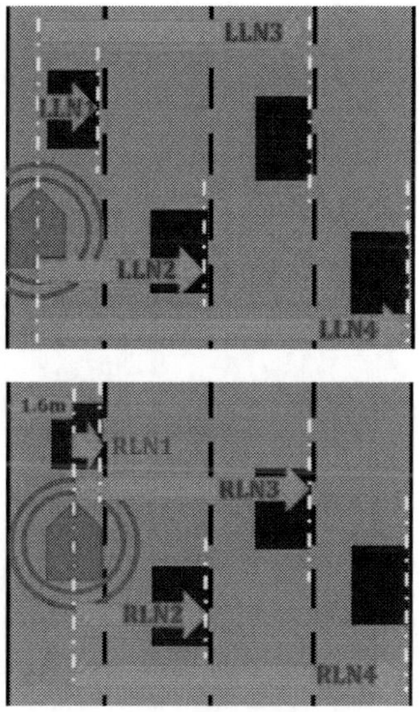

Figure 3. Lane Numbers and Parameters

Depending on the driver's driving habits, the vehicle may be driven on the left side of the road or on the right. In each case, there is a positional difference of the ego-vehicle The parameter $\boldsymbol{X_{hor}}$, suggested in the algorithm of lane estimation in the ego-vehicle is the lateral distance information between the ego-vehicle extracted from the database and the relative vehicle in the vicinity of the ego-vehicle.

$\boldsymbol{R_{wid}}$ is the width of the road, $\boldsymbol{V_{wid}}$ is the width of the vehicle, $\boldsymbol{ML_{dis}}$ is the distance from boundary of the ego-vehicle to the center, (1)**LLN1**, (2)**LLN2**, (3)**LLN3**, (4)**LLN4** are the corresponding parameters for when the vehicle is driven on the left side of the lane. (1), (2), (3), and (4) means how many lanes are present depending on the value of the distance between the ego-vehicle and the perceived relative vehicle.

(1) means that there is one lane between the ego-vehicle and the perceived relative vehicle, (2) two, (3) three, (4) four lanes. (5) **RLN1**, (6) **RLN2**, (7) **RLN3**, (8) **RLN4** are corresponding parameters for when the vehicle is driven on the right side of the road. (5), (6), (7), and (8), likewise indicate how many lanes exist between the ego-vehicle and the recognized relative vehicle, with the number of lanes as shown in the parameter name exists between the ego-vehicle and the relative vehicle around the recognized ego-vehicle.

The width of the road, LN1,2,3,4 was constructed by using the distance of the number of lanes, and if the relative vehicles are located at the width of the lane classified according to the number of lanes, the roadway on which the ego-vehicle is driven can be estimated. The road width was set to any value because the exact road information for the data collection section was not known, the parameters presented are expressed in Figure 3.

TABLE Ⅰ. ESTIMATION ALGORITHM FOR EACH LANE

Road type \ Estimated Lane		Estimation Algorithm	
		Ego-vehicle Left	Ego-vehicle Right
Four Lane Road	1st Lane	$-LLN3 < x_{hor} < -LLN4$ && $(-LLN2 < x_{hor} < -LLN3$ $\parallel$ $-LLN1 < x_{hor} < -LLN2)$	$-RLN3 < x_{hor} < -RLN4$ && $(-RLN2 < x_{hor} < -RLN3$ $\parallel$ $-RLN1 < x_{hor} < -RLN2)$
	2nd Lane	$-LLN2 < x_{hor} < -LLN3$ && $(-LLN1 < x_{hor} < -LLN2$ $\parallel$ $RLN1 < x_{hor} < RLN2)$	$-RLN2 < x_{hor} < -RLN3$ && $(-RLN1 < x_{hor} < -RLN2$ $\parallel$ $LLN1 < x_{hor} < LLN2)$
	3rd Lane	$RLN2 < x_{hor} < RLN3$ && $(RLN1 < x_{hor} < RLN2$ $\parallel$ $-LLN1 < x_{hor} < -LLN2)$	$LLN2 < x_{hor} < LLN3$ && $(LLN1 < x_{hor} < LLN2$ $\parallel$ $-RLN1 < x_{hor} < -RLN2)$
	4th Lane	$LLN3 < x_{hor} < LLN4$ && $(LLN2 < x_{hor} < LLN3$ $\parallel$ $LLN1 < x_{hor} < LLN2)$	$RLN3 < x_{hor} < RLN4$ && $(RLN2 < x_{hor} < RLN3$ $\parallel$ $RLN1 < x_{hor} < RLN2)$
Three Lane Road	1st Lane	$-LLN2 < x_{hor} < -LLN3$ && $-LLN1 < x_{hor} < -LLN2$	$-RLN2 < x_{hor} < -RLN3$ && $-RLN1 < x_{hor} < -RLN2$
	2nd Lane	$-LLN1 < x_{hor} < -LLN2$ && $RLN1 < x_{hor} < RLN2$	$-RLN1 < x_{hor} < -RLN2$ && $LLN1 < x_{hor} < LLN2$
	3rd Lane	$RLN2 < x_{hor} < RLN3$ && $RLN1 < x_{hor} < RLN2$	$LLN2 < x_{hor} < LLN3$ && $LLN1 < x_{hor} < LLN2$
Two Lane Road	1st Lane	$-LLN1 < x_{hor} < -LLN2$	$-RLN1 < x_{hor} < -RLN2$
	2nd Lane	$RLN1 < x_{hor} < RLN2$	$LLN1 < x_{hor} < LLN2$
One Lane Road	1st Lane	$-LLN1 < x_{hor} < RLN1$	$-RLN1 < x_{hor} < LLN1$

A lane width section of 1st, 2nd, 3rd, 4th means the number of lanes existing between the ego-vehicle and recognized relative vehicle. The estimated result of the algorithm depends on how many lanes are existed between the ego-vehicle and the recognized relative vehicle. expressed in TABLE Ⅰ. The method of applying the extracted relative vehicle information to algorithm will be explained after looking at a Figure 4.

The process of estimating a lane is presented in Figure 4. First extract the lateral distance coordinates of the recognized relative vehicle around the ego-vehicle from the database. Convert the extracted coordinates to the distance value of the (meter) unit and apply them to the algorithm. Then, it is necessary to verify that the distance values of the relative vehicles around the ego-vehicle were in the lane width section given in the algorithm. If the relative vehicle is recognized in the road width section of the algorithm, then the lane of the self-vehicle is estimated, but if it is not recognized, it is reapplied to the algorithm or cannot be estimated to lane of ego-vehicle. The data applied to the algorithm was collected from the reference vehicle (Figure 1) and brought to the file format. It also provided a viewer program that allows users to view data from the Spatial Information Research Institute, which has collected sensor data. Therefore, information such as distance, coordinates, etc. regarding objects around the ego-vehicles is provided in the form of CVS files. Currently, data is applied offline to the algorithm and automatically performed to estimate the position of the ego-vehicle.

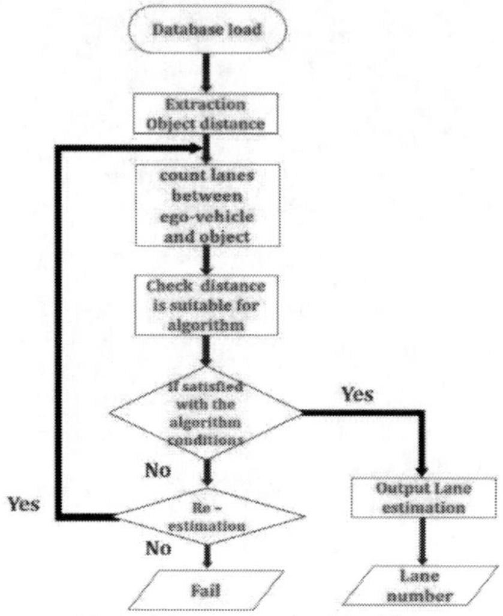

Figure 4. Lane estimation Algorithm

Ⅳ. ALGORITHM VERIFICATION RESULTS

For performance evaluation of developed algorithm, curved roads, tunnels, 1,2,3,4 lane roads, overpass and roads with curvature sections were extracted from the database and shown in Figure 5.

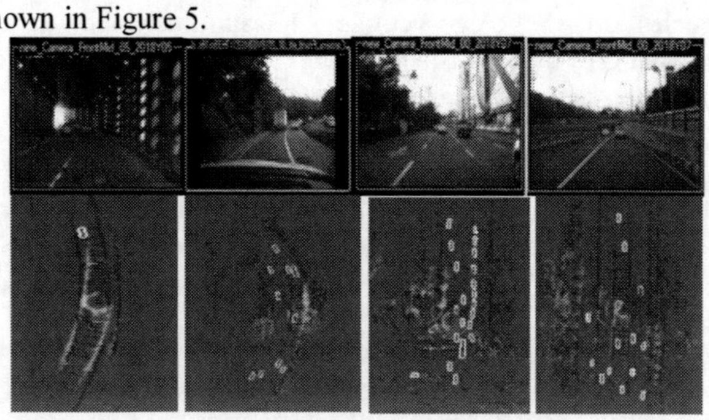
Figure 5. Tunnel, Curve, Overpass, Multi lane Road

The performance evaluation was conducted by comparing the result of lane estimation algorithm for vehicles developed in this study with actual driving data. The evaluation environment was established with 71 specific point-in-time driving data sections extracted from the database and eight types of roads. For each section of the driving data, the accuracy was shown by comparing the actual ego-vehicle lane to the lane estimated by algorithm.

For example, an evaluation of performance at a specific point in time shows that the lane of the ego-vehicle estimated by the algorithm is second lane and coincides with the lane of the ego-vehicle actually located when compared with the actual driving data. Such fact can be found in Figure 6. By the above evaluation method estimated the lane from 71 actual driving data specific points-in-time.

As a result of the estimation, the accuracy of the algorithm is divided into Known, which knows the total number of lanes on the road, and unKown, which has no information on the number of lanes. TABLE Ⅲ shows the accuracy of whether the lane estimated by the vehicle algorithm matches the lane in which the vehicle is located in the actual driving data.

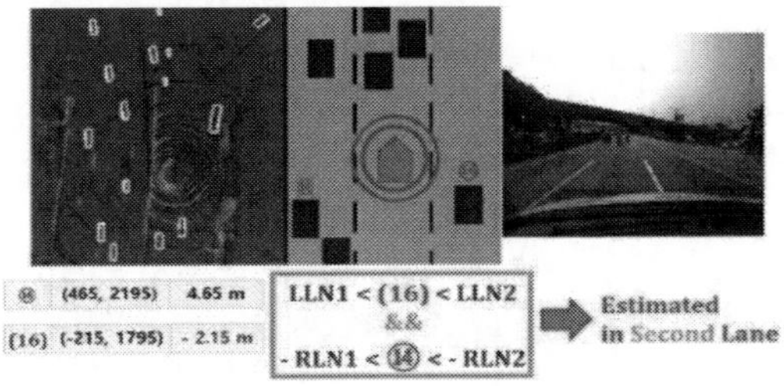

Figure 6. Lane estimation result from algorithm at a specific section in database

TABLE Ⅱ. PARAMETER NAME & VALUE

Parameter \ Value	Ego-Vehicle Left	Ego-Vehicle Right
V_{wid}	1.9 m	
R_{wid}	3.5 m	
ML_{dis}	0.95 m	
LLN1 \ RLN1	2.55 m	0.95 m
LLN2 \ RLN2	6.05 m	4.45 m
LLN3 \ RLN3	9.55 m	7.95 m
LLN4 \ RLN4	13.05 m	11.45 m

TABLE Ⅲ. LANE ESTIMATION ALGORITHM ACCURACY

Position \ Classification	KNOWN	UNKNOWN
	Accuracy [%]	
1st Lane	19/20 [95 %]	13/20 [65 %]
2st Lane	20/22 [90.9 %]	5/22 [22.72 %]
3st Lane	19/21 [90.48 %]	11/21 [52.38 %]
4st Lane	6/8 [75 %]	4/8 [50 %]
Total	90.14%	46.48%

V. CONCLUSION

The algorithm verification of 71 actual driving data sections showed high accuracy, with **90.14%** case knowing the total number of lanes and **46.48%** not knowing the total number. Therefore, if there were no dynamic obstacles on top of all lanes, it was not possible to distinguish whether the lane on which the ego-vehicle is located was a second or a third. The reason for the decrease in unknown accuracy is that if the total number of total lanes are not known, objects on other roads can be recognized. This satisfies the various conditions of the algorithm and results in multiple lanes. However, if one continues to build previous lane data to the time axis, it can distinguish this case unless ego-vehicle is rapidly changing lane. Using the estimated lane information from the previous lanes of the stacked data every hour, the estimated results will be much better in any case. This research is still a work in progress, but in the future studies we will use a more flexible and accurate kalman filter or particle filter than a specific condition-based algorithm, The estimation algorithm with these filters will also be able to estimate the curve roads on which the curvature exists. Finally, the autonomous vehicle "**D2**" held by the Autonomous a2z will be used to collect data directly from specific sections and apply the algorithm developed to the actual vehicles to verify the performance of algorithm in real time driving.

ACKNOWLEDGEMENT

This work was supported by the Technology Innovation Program (10079730, Development and Evaluation of Automated Driving Systems for Motorway and City Road and driving environment) funded By the Ministry of Trade, Industry & Energy (MOTIE, Korea).

REFERENCES

[1] Seo.J and Jeong .H, "Precision Vehicle Location System Based on Sensor Fusion and Self-Driving Technology," Automation Systems, Tech Insight, pp. 42-47, 2015.

[2] Eirini.T, Surya.P, John.S and Baras, "Interest, energy and physical-aware coalition formation and resource allocation in smart IoT applications," In 2017 51st Annual Conference onInformation Sciences and Systems (CISS), pp. 1-6, 2017.

[3] Park.S, Lee.H, Jeong.Y, Kim.S, Lee,H and Yi.K, "Lane and Curb Detection based Vehicle Localization Algorithm for an Automated Driving in Complex Urban Roads," SNU Department of Mechanical & Aerospace Engineering and Institute of Advanced Machines and Design, KSME17-Th12C004, 2017.

[4] Seo.J, Jang.J, Min.D and Jeong .H, "Sensor Fusion-based Precise Vehicle Localization System," Research Institute of Automotive Control and Electronics, HanyangUniversity, WITHROBOTInc, KSAE15-A0230, 2015

A New Model for Hard Braking Vehicles and Collision Avoiding Trajectories

Fynn Terhar

BMW Group
Department for Fleet Intelligence
Munich, Germany
Email: fynn.terhar@bmw.de

Christian Icking

FernUniversität in Hagen
Department for Cooperative Systems
Hagen, Germany
Email: christian.icking@fernuni-hagen.de

Abstract—**In this paper, we propose a new model to describe vehicle dynamics in full braking situations with collision avoiding motions. By combining the equations of the classic Ackermann-Model with conditions that ensure a stable vehicle movement during simultaneous heavy braking and turning motions, we derive a model that describes the set of controllable trajectories. We describe these trajectories by compound motion equations in the x, y plane that are directly computable. We discuss our model regarding uncertainties and their effect on reachability analysis of vehicles in admissible scenarios, to show the feasibility of our solution. We compare our model to the well known Constant-Turn-Rate-And-Acceleration-Model which is computationally more expensive and less precise. By considering uncertainties of the parameters used in our model, we show a way to estimate the reachable area of a hard braking vehicle.**

Keywords–Reachability; Trajectory; Dynamic Vehicle Model; Safety; Collision Avoidance; Braking.

I. INTRODUCTION

A. Motivation

Many functions in Highly Automated Driving (HAD) and Advanced Driving Assistance Systems (ADAS) are discussed regarding their safety towards events caused by other traffic participants, whose behavior is not well predictable. In case of an unforeseen event, vehicles need to avoid a collision by a suitable trajectory. In literature, these trajectories are often referred to as Fail-Safe-Trajectories. These trajectories can either be evasive and try to find a solution around an obstacle or bring the vehicle to an emergency stop. The vehicle is then forced to find a trajectory till full stop within an area in front of the vehicle, which is defined by its physical properties and speed vector. In this paper, we call this area the *braking area*, which is important to know in many different applications. For example, when defining the set up of on-board sensors, it can be useful to have a good knowledge of the braking area. Also when searching for fail-safe trajectories, the knowledge of the reachable set of vehicle states can significantly accelerate the computation, as it reduces the search space and can therefore save valuable time in emergency situations.

B. Literature overview

Computing the braking area of a vehicle is related to finding fail-safe trajectories. Methods for avoiding obstacles are numerous, see for example Werling et al. [1], where the authors address dynamic street scenarios by an optimal control approach. The method generates trajectories that are optimal in terms of jerk minimization and following a previously computed trajectory. Another approach is explained by Ziegler et al. [2]. They use a cost function to plan obstacle avoiding paths in unstructured environments, but not on the description of fail-safe trajectories. Several approaches towards finding fail safe trajectories for road vehicles exist. Pek and Althoff [3] describe a method to generate fail-safe trajectories for dynamic traffic scenarios in a computationally efficient manner. Their solution approximates the set of reachable states of the ego vehicle and other traffic participants and can therefore guarantee collision free trajectories. A motion planner for fail-safe trajectories is shown by Magdici and Althoff [4]. A related application is presented in [5], where a safety framework is demonstrated that can test a planned trajectory for possible future collisions.

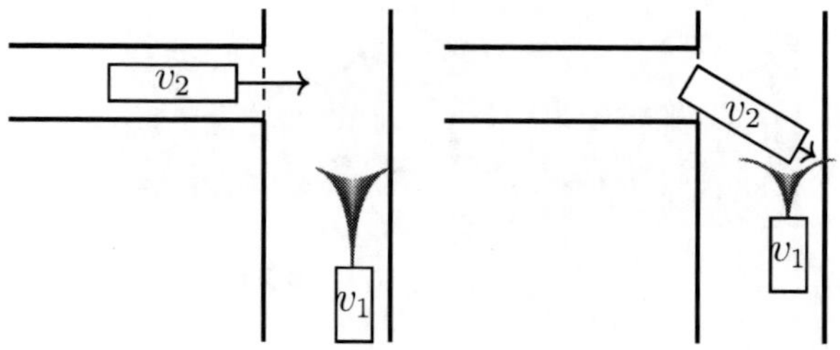

Figure 1. A critical traffic situation. Left, two vehicles approach a T-crossing without seeing each other. Right, vehicle v_1 is suddenly confronted with the long vehicle v_2 which blocks the road. To remain safe, v_1 should always know its reachable area in case of emergency braking.

Mitchell et al. [6] discuss different approaches of reachability analysis of dynamic systems for the safety assessment of trajectories. Asarin et al. [7] present an approach for reachability approximation of partially linearized systems in general. An often applied technique to approximate the state space efficiently is by zonotopes, see, e.g., the paper of Girad [8]. Koschi et al. [9] introduce an open source software solution which predicts road occupancy by traffic participants within a given time horizon. By overestimating the occupancy by the union of several object models, the authors ensure to find all possible traffic configurations. Potential braking and turning is overestimated by a circle of lateral and longitudinal maximum and minimum accelerations. The physical interaction between velocity and admissible lateral accelerations are therefore overestimated. Althoff [10] describes many underlying concepts of reachability analysis for road vehicles. In contrast to formal verification, ByeoungDo et al. [11] propose a Recurrent Neural Net for predicting traffic participants. Explicit braking and turning motions and their interrelation are not in the focus.

Our model provides a more detailed and accurate description of this interaction in order to reduce the overestimation towards a more realistic model.

The interrelation of braking and turning is, e.g., discussed by Giovannini et al. [12] where the authors describe the last point in time when a collision can be avoided by swerving. The authors explicitly focus their work on two-wheeled vehicles. Ackermann et al.[13] present control strategies for braking and swerving motions. Choi et al. [14] propose an additional strategy based on model predictive control.

C. Contribution

In this work, we present an accurate model for estimating not only a set of feasible trajectories of a vehicle while braking and turning till full stop. We also discuss the model regarding parameter uncertainties, to describe their effect on the overall *braking area*. Thereby, we aim to overestimate the occupancy where necessary, while reducing it where it is possible in order to provide both safety and accuracy.

In Section II, we define a model that directly calculates vehicle trajectories towards a full stop while simultaneously braking and steering. Braking and steering always needs to be done in a balanced way, as both influence the controllability of the vehicle on the road. We therefore introduce a parameter that describes the ratio of this compromise. Furthermore, the friction between different road surfaces and tires is considered, as well as the vehicles' dynamic limits and initial state. In Section III, we discuss how uncertainties of the model parameters influence the braking area. We thereby provide an estimation of the braking area in admissible situations.

II. Model definition

Physical model values are denoted as regular latin letters, while angles are denoted as greek letters. Symbols used in this paper are summarized in the following Table I:

TABLE I. SYMBOLS USED IN THIS PAPER.

Symbol	*Description*	*Unit*
X_i	Model state at time i	–
p	Position $\in \mathbb{R}^2$	m
X_{stop}	Stop state, $v = 0$	–
ψ	Yaw Angle	rad
b	Braking Factor	–
$\hat{a}$	Maximum admissible acceleration	m/s^2
r_{turn}	Minimum turning radius	m
$\mathcal{I}_\bullet$	Interval of admissible values for $\bullet$	–
$\bullet_{\min}, \bullet_{\max}$	Extreme values of $\mathcal{I}_\bullet$	–
$\underline{f}(t), \overline{f}(t)$	Lower/Upper part of function $f(t)$	–

A. Assumptions

Our model builds upon assumptions that describe braking and turning in the following order of priority:

1) The vehicle needs to stop as quickly as possible.
2) By steering, the vehicle must try to avoid obstacles if ever possible, or minimize an unavoidable impact.

These assumptions hold in many situations where a collision can only be prevented by a full stop of the vehicle, as sketched out in Figure 1. Due to the high speed of v_2, the other vehicle v_1 can only react to the depicted incident by quickly stopping. The purpose of the following model is to predict the set of possible trajectories in space and time during braking maneuvers in such situations.

B. Model derivation

Our model builds on the so called *friction circle*, e.g. described by Pacejka [15]. As the modeled vehicle is braking in order to come to a full stop quickly, it will always be located near the boundary of this circle, either due to braking only, or by braking and turning in combination. Staying at the boundary of the friction circle means that the vehicle remains controllable in such an extreme maneuver. The basic concept of the friction circle is shown in Figure 2. The combined acceleration a_{res} is the vectorial sum of the centripetal acceleration $a_{\text{cen}} = v\underline{\dot{\psi}}$ and the longitudinal acceleration a_{lon}. An $a_{\text{res}} > \hat{a}$ can not be achieved, because the tires would loose their grip.

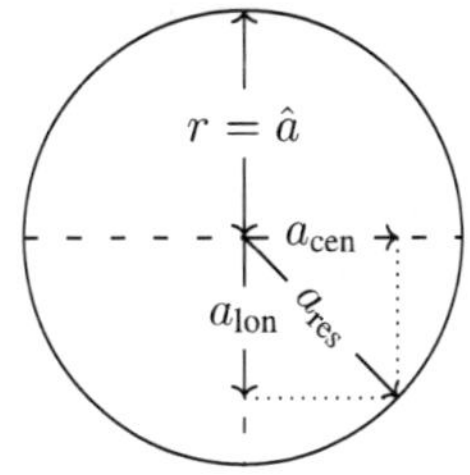

Figure 2. Friction circle in the a_x, a_y-plane. Radius r is equal to the maximally applicable acceleration $\hat{a}$ between vehicle and road surface.

Note, that the friction circle as shown in Figure 2 is an idealized and simplified model of tire forces. A more accurate model like the friction ellipse [15] will be implicitly considered in the reachability estimation in Section III, by introducing a high uncertainty in $\hat{a}$. The circle leads to the following equation:

$$a_{\text{res}} = \sqrt{a_{\text{cen}}^2 + a_{\text{lon}}^2} = \sqrt{(v\underline{\dot{\psi}})^2 + a_{\text{lon}}^2} \tag{1}$$

For a braking and turning maneuver in an emergency situation, we want to keep the vehicle controllable but also apply the strongest acceleration possible in order to react effectively. This constraints the vehicle to operate on the boundary of the friction circle, as described by (2).

$$\hat{a} = \sqrt{(v\underline{\dot{\psi}})^2 + a_{\text{lon}}^2} \tag{2}$$

1) Interrelation between braking, steering and yaw rate: The yaw rate $\dot{\psi}$ describes the change in yaw angle ψ of a vehicle over time. As the acceleration a_{res} results from a combination of braking and steering, the ratio $a_{\text{lon}}/\hat{a}$ causes different trajectories. We define this ratio by the factor b, as declared in (3), further on called *Braking Factor*. We call b Braking Factor, as it describes the percentage of $\hat{a}$ that is applied for braking rather than turning. A b value of -0.5 means that 50% of the applicable acceleration is applied for braking. Note, that $\hat{a}$ is positive, but when braking a_{lon} is negative, hence we choose $b \in [-1, 0]$.

$$b := \frac{a_{\text{lon}}}{\hat{a}} \tag{3}$$

Solving (2) for $\underline{\dot{\psi}}$ yields a description of $\underline{\dot{\psi}}(t)$, see (4).

$$\underline{\dot{\psi}}(t) = \frac{\hat{a}\sqrt{1 - b^2}}{v(t)} \tag{4}$$

Equation (4) describes the yaw rate $\dot{\underline{\psi}}(t)$ of a vehicle that stays at the boundary of the friction circle with radius $\hat{a}$. This describes the interrelation between braking, steering, and yaw rate.

2) Vehicle yaw angle as function of time: The vehicle's yaw angle ψ determines its travel direction, so a description of $\underline{\psi}(t)$ is required for the model, as shown in (5).

$$\underline{\psi}(t) = \int \dot{\underline{\psi}}(t)dt = Z(\ln(v(t)) - \ln(v_0)) + \psi_0 \tag{5}$$

$$v(t) = a_{\text{lon}}t + v_0 \tag{6}$$

$$Z = b^{-1}\sqrt{1-b^2} \tag{7}$$

where $v(t)$ is the linear speed equation (6) and Z is a constant described by (7).

The angle $\underline{\psi}(t)$ rises in its absolute as the speed v falls. Figure 3 depicts this relation for an exemplary vehicle. The major flaw of this description is that the yaw rate tends towards ∞. This is not possible for any real vehicle, as the limit for a real vehicle is reached when the steering wheel reaches its maximum position and the minimum turning angle is performed. This effect is depicted in Figure 3 in the dashed line. As the speed approaches zero, $\underline{\psi}(t)$ approaches ∞. A realistic model must therefore respect the smallest *turning radius* r_{turn}. As $\dot{\psi}$ of a moving object is also defined as

$$\dot{\psi}(t) = \frac{v(t)}{r}$$

where r is the radius of the object's circular path, the maximal $\dot{\psi}(t)$ can be described by (8).

$$\dot{\overline{\psi}}(t) = \frac{v(t)}{r_{\text{turn}}} \tag{8}$$

By solving

$$\frac{\hat{a}\sqrt{1-b^2}}{a_{\text{lon}}t + v_0} = \frac{v(t)}{r_{\text{turn}}}$$

for t we know the time t_{crit} at which the yaw rate will reach its mechanical maximum, as shown in (9).

$$t_{\text{crit}} = \frac{\sqrt{r_{\text{turn}}\hat{a}\sqrt{1-b^2}} - v_0}{a_{\text{lon}}} \tag{9}$$

At times above t_{crit} we therefore describe the yaw angle by $\overline{\psi} = \int \dot{\overline{\psi}}(t)dt$, as shown in (10), in order to derive a realistic model.

$$\overline{\psi}(t) = \int \frac{v(t)}{r_{\text{turn}}}dt = \frac{\frac{1}{2}a_{\text{lon}}t^2 + v_0t + const_{\overline{\psi}}}{r_{\text{turn}}} \tag{10}$$

where $const_{\overline{\psi}}$ must be defined in a way that the condition

$$\underline{\psi}(t_{\text{crit}}) = \overline{\psi}(t_{\text{crit}}) \tag{11}$$

holds. The condition means that the angle at t_{crit} must be equal for both (5) and (10). It yields $const_{\overline{\psi}}$ as:

$$const_{\overline{\psi}} = r_{\text{turn}}\underline{\psi}(t_{\text{crit}}) - \frac{1}{2}a_{\text{lon}}t^2 - v_0t \tag{12}$$

The sectionally defined yaw angle ψ, consisting of $\underline{\psi}$ and $\overline{\psi}$, is plotted in Figure 3 (solid line). Note, how ψ now drops with falling speed, which directly follows from (8). The dashed line plots $\underline{\psi}(t)$, which approaches ∞ as the speed approaches zero. This follows from its property to be at the boundary of the friction circle. At low speeds, this can only be achieved by high yaw rates.

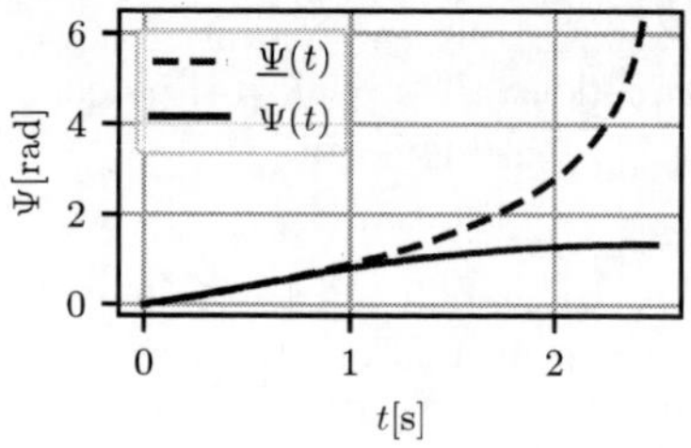

Figure 3. The yaw angle ψ over time t during braking and steering. Dashed line plots $\underline{\psi}(t)$. Solid line plots the combined stepwise definition $\psi(t)$ which considers the turning radius r_{turn} for $t > t_{\text{crit}}$.

The final description of $\psi(t)$ is defined stepwise in (13).

$$\psi(t) = \begin{cases} \underline{\psi}(t), & 0 \le t \le t_{\text{crit}} \\ \overline{\psi}(t), & t_{\text{crit}} < t \le t_{\text{stop}} \end{cases} \tag{13}$$

Note, that due to the equality condition of yaw angles in (11) and the definition of t_{crit} in (9) the final yaw angle $\psi(t)$ is differentiable. Also note, that then $t > t_{\text{crit}}$, the vehicle in our model is no longer at the boundary of the friction circle.

3) Vehicle position as function of time: A description of vehicle position $p(t) = [x, y]$ is described as the compound equations for x and y, which follow from the integrals:

$$x(t) = \int v(t)\cos(\psi(t))dt$$

$$y(t) = \int v(t)\sin(\psi(t))dt$$

Solving the integrals yields:

$$\underline{x}(t) = \frac{v(t)^2\left(Z\sin(\psi(t)) + 2\cos(\psi(t))\right)}{a_{\text{lon}}\left(Z^2+4\right)} + C_{\underline{x}} \tag{14}$$

$$\overline{x}(t) = r_{\text{turn}}\sin(\psi(t)) + C_{\overline{x}} \tag{15}$$

These equations describe position over time $x(t)$ and $y(t)$. See stepwise (16) for $x(t)$.

$$x(t) = \begin{cases} \underline{x}(t), & 0 \le t \le t_{\text{crit}} \\ \overline{x}(t), & t_{\text{crit}} < t \le t_{\text{stop}} \end{cases} \tag{16}$$

The constant $C_{\underline{x}}$ is bound by the conditions $x(0) = x_0$, which means the vehicle must be at the starting position at time t_0. The constant for $\overline{x}$, $C_{\underline{x}}$ is bound to hold the condition $\overline{x}(t_{\text{crit}}) = \underline{x}(t_{\text{crit}})$, which means that $\underline{x}$ must seamlessly – e.g. in value and gradient – be continued by $\overline{x}$ at t_{crit}. The result for both constants is described by (17) and (18).

$$C_{\underline{x}} = x_0 - \frac{v_0^2\left(Z\sin(\psi_0) + 2\cos(\psi_0)\right)}{a_{\text{lon}}(Z^2+4)} \tag{17}$$

$$C_{\overline{x}} = x(t_{\text{crit}}) - r_{\text{turn}}\sin(\psi(t_{\text{crit}})) \tag{18}$$

The general description for $y(t)$ is shown below in (21), and can be derived analogously to $x(t)$.

$$\underline{y}(t) = -\frac{v(t)^2\,(Z\cos(\psi(t)) - 2\sin(\psi(t)))}{a_{\text{lon}}\,(Z^2+4)} + C_{\underline{y}} \tag{19}$$

$$\overline{y}(t) = -r_{\text{turn}}\cos(\psi(t)) + C_{\overline{y}} \tag{20}$$

$$y(t) = \begin{cases} \underline{y}(t), & 0 \le t \le t_{\text{crit}} \\ \overline{y}(t), & t_{\text{crit}} < t \le t_{\text{stop}} \end{cases} \tag{21}$$

$$C_{\underline{y}} = y_0 - \frac{v_0^2\,(Z\cos(\psi_0) - 2\sin(\psi_0))}{a_{\text{lon}}(Z^2+4)}$$

$$C_{\overline{y}} = y(t_{\text{crit}}) + r_{\text{turn}}\cos(\psi(t_{\text{crit}}))$$

The trajectory of a braking and turning vehicle is described as $p(t)$, by the compound x- and y-position in Cartesian coordinates over time. How the more realistic yaw angle description influences the resulting position can be seen in a direct comparison in Figure 4. The vehicle performs a spiral shape until the maximum turning angle is reached, which is clearly visible in Figure 4b. In a real situation, this trajectory with such a low b value will most likely not be considered feasible for braking, it rather demonstrated the spiral nature of our model. Note, that all other $b \in \,]-1, 0[$ also describe spirals, only less clearly visible as in Figure 4a.

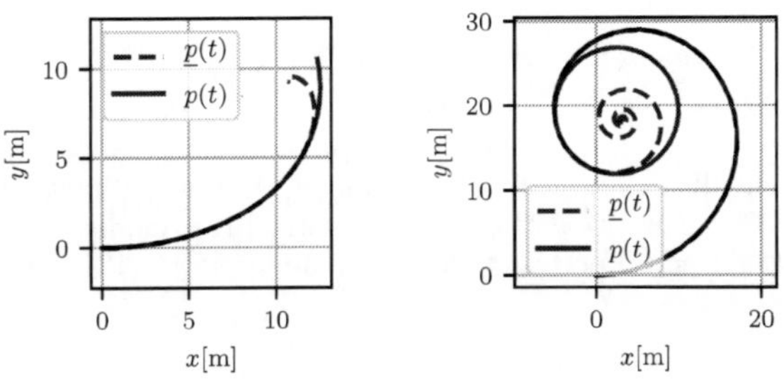

(a) Calculated position $p(t)$ with $b = -0.5$.

(b) Calculated position with $b = -0.085$.

Figure 4. Vehicle position $p(t)$ in x, y plane with different values for b. Dashed line, the model result without considering r_{turn}. Solid line, the model considering r_{turn}, using the final model equations.

In the next step, we compare our trajectories to simulative results of another model.

4) Comparison of our model against CTRA model: To evaluate our model's performance with respect to calculation time and to show its correctness, we compare it to a CTRA-model [16] *Constant Turn Rate and Acceleration* in a simulation. The CTRA simulation iteratively moves a vehicle, such that our condition in (2) is fulfilled, and the assumptions introduced in Section II hold. The simulation therefore calculates effectively the same maneuvers as our model, but in a very different way. We choose the CTRA-model, as it is well known, allows the vehicle to follow a spiral shape and has the same state space representation as our model. The turn rate and acceleration is assumed to be constant within one of many consecutive time steps Δt.

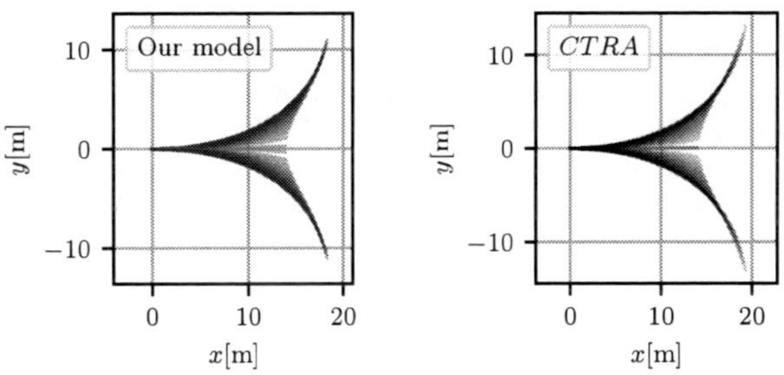

(a) Our braking model.

(b) CTRA model, $\Delta t = 0.0075s$.

Figure 5. Comparison of our model to the CTRA model for 40 vehicle trajectories with linearly sampled b values. The starting conditions for both tests are $v_0 = 16.67 m/s, \hat{a} = 10 m/s^2, r_{\text{turn}} = 12.5m, \psi_0 = 0\ rad$.

The result in Figure 5 shows that our model matches the shape of the CTRA-model well, without introducing linearization errors as the CTRA model does.

Both results from Figure 5 show a very similar structure. Note that the CTRA model (Figure 5b) has slightly longer trajectories, especially in the outer arms of the structure. This is caused by the CTRA-model's assumption of a constant turn rate $\dot{\psi}$, which is not correct in this kind of non-linear maneuver. In our model (Figure 5a), the only assumption is that of a constant acceleration, as introduced in Section II.

The main advantage of our model is the fact that we can directly compute certain vehicle positions straight from the formulas derived in Section II such that time intensive calculations are not necessary. A comparison of computation times t_{calc} in seconds, and their deviation $\sigma_{t_{\text{calc}}}$ over 10 runs is shown in Table II. In the first test, only the stop states where computed of 1000 different b values. In the second test, a whole pearl chain of positions from start to stop was computed, with 250 points per b value.

TABLE II. COMPARISON TO THE CTRA MODEL.

Calculate 1000 possible **stop states**, $\Delta t = 0.01112s$						
v_0	5 m/s		10 m/s		20 m/s	
	Mean t_{calc}[s]	$\sigma_{t_{\text{calc}}}$	Mean t_{calc}	$\sigma_{t_{\text{calc}}}$	Mean t_{calc}	$\sigma_{t_{\text{calc}}}$
CTRA	1.0715	0.0137	2.1975	0.0052	4.9310	0.1073
Our model	0.2059	0.0053	0.2078	0.0017	0.2144	0.0075
Calculate 1000 **trajectories**, 250 samples per trajectory, $\Delta t = 0.01112s$						
v_0	5 m/s		10 m/s		20 m/s	
	Mean t_{calc}	$\sigma_{t_{\text{calc}}}$	Mean t_{calc}	$\sigma_{t_{\text{calc}}}$	Mean t_{calc}	$\sigma_{t_{\text{calc}}}$
CTRA	1.0870	0.0207	2.2335	0.0096	4.9761	0.0814
Our model	0.2310	0.0017	0.2326	0.0021	0.2320	0.0011

The table shows that our model is up to 20 times faster in terms of computing time than the CTRA model, especially for high initial velocities v_0. This is caused by the fact that CTRA must iteratively compute time steps until the stop position is found, whereas our model can directly compute the stop state.

III. DISCUSSION OF MODEL UNCERTAINTIES

In this section, we discuss the effect of individual uncertainties in the model parameters $r_{\text{turn}}, \hat{a}$ and the initial vehicle state $X_0 = [x_0, y_0, v_0, \psi_0]^T$. We model the uncertainties as intervals $\mathcal{I}_\Theta, \mathcal{I}_{X_0}$ that contain all possible values.

A. Highest possible deceleration $\hat{a}$

The highest possible deceleration heavily depends on the road and tire conditions, which are often uncertain. The interval $\mathcal{I}_{\hat{a}}$ therefore covers the most slippery and most rough road condition possible. Calculating different stop states X_{stop} with different values for $\hat{a}$ reveals an almost linear behavior within expectable values of $\hat{a} \in \mathcal{I}_{\hat{a}}$.

The resulting shape of 50 different $\hat{a} \in \mathcal{I}_{\hat{a}}$ can be seen in Figure 6a, where lower values of $\hat{a}$ lead to a farther vehicle trajectory with an almost linear behavior.

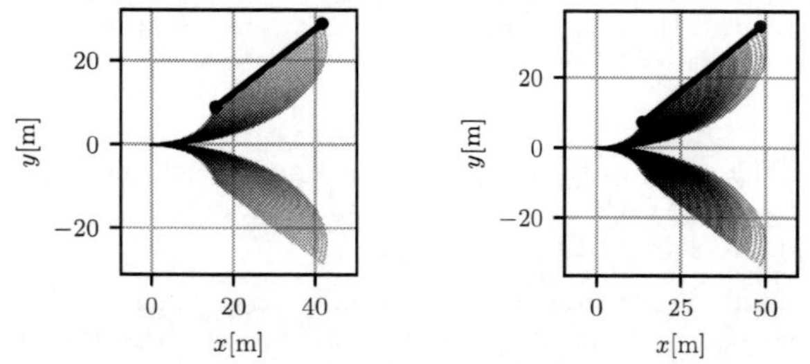

(a) Resulting trajectories at interval $\mathcal{I}_{\hat{a}} = [4, 12]\,{}^{m}/_{s^2}$.

(b) Resulting trajectories at intervals $\mathcal{I}_{\hat{a}} = [4, 12]\,{}^{m}/_{s^2}$, $\mathcal{I}_{v_0} = [15.3, 18.1]\,{}^{m}/_{s}$

Figure 6. Two sets of trajectories with a b value of -0.6. Left, only considering $\mathcal{I}_{\hat{a}}$. Right, considering $\mathcal{I}_{\hat{a}}$ and $\mathcal{I}_{v_0}$. A line segment shows the extending effect of the parameter uncertainties on the top half.

B. Smallest possible turning radius r_{turn}

The smallest possible turning radius r_{turn} is a vehicle inherent parameter which influences the trajectory after t_{crit} and also defines the value of t_{crit} itself. Although there are certain legal requirements for r_{turn} depending on vehicle class, the exact value is uncertain, especially when considering other traffic participants.

Any $r_{\text{turn}} \in \mathcal{I}_{r_{\text{turn}}}$ causes a different stopping position. Unfortunately, the lowest or highest $r_{\min}$ not always leads to the outmost stopping position. By observing the stopping positions depending on r_{turn}, one can see that the shape of all stopping positions with different $r_{\text{turn}} \in \mathcal{I}_{r_{\text{turn}}}$ forms a spiral with a rising radius. Let A be the stopping position of the lowest r_{turn}, $A = X_{\text{stop}}|_{r_{turn,min}}$, and $B = X_{\text{stop}}|_{r_{turn,max}}$. The circle with radius $r = dist(A, B)$ at center A then includes all points of the spiral, which means all stopping positions can be overestimated by such a circle. By describing this distance as function $d = f(\hat{a}, v_0)$, it can be shown that the maximum distance is at $d_{max} = f(\hat{a}_{min}, v_{0,max})$. Figure 7 shows an example of such a circle.

In order to show the spiral effect in Figure 7, we assumed $\mathcal{I}_{r_{\text{turn}}} = [1e{-}7, 13]m$ and $v_0 = 10\,{}^{m}/_{s}$, which results in a circle radius of $\approx 2.4m$. For a more realistic scenario of $\mathcal{I}_{r_{\text{turn}}} = [7, 13]m$ and $v_0 = 10\,{}^{m}/_{s}$, the radius of the circle is $\approx 1.3m$.

C. Initial velocity v_0

The uncertainty in the initial velocity $\mathcal{I}_{v_0}$ determines the stopping distance similarly to $\mathcal{I}_{\hat{a}}$, as it stretches the possibly reachable positions farther from the start. This means the closest reachable position is defined by $v_{0,min}$ and $\hat{a}_{max}$, which stands for a very rough road-to-tire surface. In contrast, the farthest reachable stopping position is defined by the highest velocity $v_{0,max}$ on the most slippery road $\hat{a}_{min}$ possible. An example of the resulting shape is shown in Figure 6b.

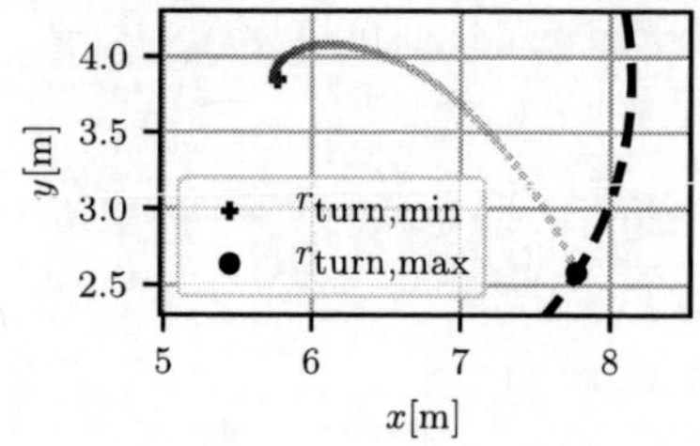

Figure 7. Effect of $\mathcal{I}_{r_{\text{turn}}}$ on X_{stop}. The figure shows how a circle can surround all stopping positions caused by different $r_{\text{turn}} \in \mathcal{I}_{r_{\text{turn}}} = [1e{-}7, 13]m$.

D. Initial position

The initial position of the vehicle will always be uncertain, as no perfect localization is possible. The effect of an uncertain starting position (x_0, y_0) is however not complex, as a different starting position of $\Delta x, \Delta y$ simply causes a translation of the complete reachable area of $\Delta x, \Delta y$.

E. Initial yaw angle

The initial yaw angle rotates the complete reachable area around the starting position of the vehicle. Figure 8a shows an example of this effect, where $\mathcal{I}_{\psi_0} = [-{}^{\pi}/_{32}, {}^{\pi}/_{32}]$.

F. Combination of all uncertainties

So far, we discussed the uncertainty of parameters separately. To describe and overestimate all system states that can potentially be reached under all uncertainties is not in the scope of this paper. In order to do so, a formal reachability analysis must be performed, compare for example [5][6][10][17].

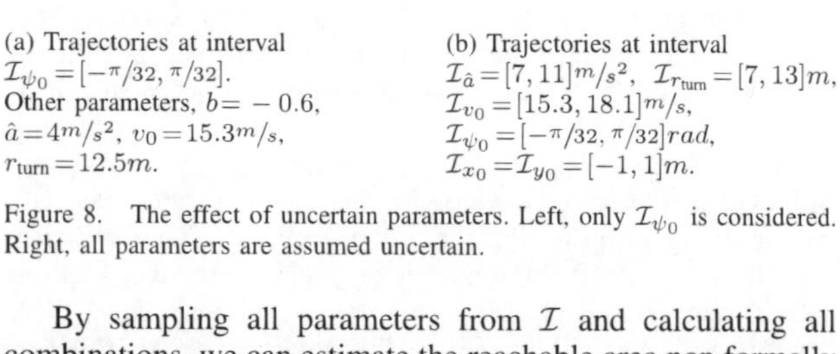

(a) Trajectories at interval $\mathcal{I}_{\psi_0} = [-{}^{\pi}/_{32}, {}^{\pi}/_{32}]$. Other parameters, $b = -0.6$, $\hat{a} = 4\,{}^{m}/_{s^2}$, $v_0 = 15.3\,{}^{m}/_{s}$, $r_{\text{turn}} = 12.5m$.

(b) Trajectories at interval $\mathcal{I}_{\hat{a}} = [7, 11]\,{}^{m}/_{s^2}$, $\mathcal{I}_{r_{\text{turn}}} = [7, 13]m$, $\mathcal{I}_{v_0} = [15.3, 18.1]\,{}^{m}/_{s}$, $\mathcal{I}_{\psi_0} = [-{}^{\pi}/_{32}, {}^{\pi}/_{32}]rad$, $\mathcal{I}_{x_0} = \mathcal{I}_{y_0} = [-1, 1]m$.

Figure 8. The effect of uncertain parameters. Left, only $\mathcal{I}_{\psi_0}$ is considered. Right, all parameters are assumed uncertain.

By sampling all parameters from $\mathcal{I}$ and calculating all combinations, we can estimate the reachable area non formally by the union of the resulting shapes. In Figure 8b we show such a result, where $\mathcal{I}_{\hat{a}} = [7, 11], \mathcal{I}_{r_{\text{turn}}} = [7, 13], \mathcal{I}_{v_0} = [15.3, 18.1], \mathcal{I}_{\psi_0} = [-{}^{\pi}/_{32}, {}^{\pi}/_{32}], \mathcal{I}_{x_0} = \mathcal{I}_{y_0} = [-1, 1]$. We sample 3 parameters of each interval.

IV. Conclusion

In this paper, we present a model for hard braking and collision avoiding vehicle trajectories. We take into account the maximally applicable acceleration/deceleration between tires and road surface, the minimal turning radius, the vehicle velocity, as well as starting position and heading. We explain our approach in detail and compare our model equations with an iterative CTRA-model simulation, which finds very similar solutions. However, in tests we could show that our solution computes stopping positions and trajectories up to 20 times faster than CTRA. By solving the compound differential equations for position in x, y-plane, we describe the complete vehicle motion till full stop, while also turning and still respecting the friction circle. With the derived equations, we can directly compute possible positions that a vehicle will reach in a braking and collision avoiding scenario. This might be used to generate braking and collision avoiding trajectories, by sampling our model for different feasible motion primitives, which can be computed in very short time.

We contribute a model that can aid in solving reachability problems for hard braking vehicles in an accurate and yet overapproximative way, considering all uncertainties in model parameters and start state of the vehicle.

As next steps, the proposed model for vehicle motion can be compared to the trajectories of real vehicles under the same assumptions given. Another next step might be the usage of our model for fast generation of braking trajectories by sampling motion primitives and compare the solution to other state of the art methods. As we can directly compute motion primitives for the highly non linear motions in braking and collision avoidance the proposed model can significantly reduce valuable trajectory generation time. Another aspect that can be tested is to apply our model in a formal reachability analysis for risk assessment in hard braking traffic scenarios and compare the solution to other contributions in the field of reachability analysis.

References

[1] M. Werling, J. Ziegler, S. Kammel, and S. Thrun, "Optimal trajectory generation for dynamic street scenarios in a frenet frame," in *Proceedings - IEEE International Conference on Robotics and Automation*, 06 2010, pp. 987 – 993.

[2] J. Ziegler, M. Werling, and J. Schröder, "Navigating car-like robots in unstructured environments using an obstacle sensitive cost function," in *IEEE Intelligent Vehicles Symposium, Proceedings*, 07 2008, pp. 787 – 791.

[3] C. Pek and M. Althoff, "Computationally efficient fail-safe trajectory planning for self-driving vehicles using convex optimization," in *Proc. of the IEEE Int. Conf. on Intelligent Transportation Systems*, 2018, pp. 1447–1454.

[4] S. Magdici and M. Althoff, "Fail-safe motion planning of autonomous vehicles," in *2016 IEEE 19th International Conference on Intelligent Transportation Systems (ITSC)*, Nov 2016, pp. 452–458.

[5] C. Pek, M. Koschi, and M. Althoff, "An online verification framework for motion planning of self-driving vehicles with safety guarantees," in *AAET - Automatisiertes und vernetztes Fahren*, 01 2019, pp. 260–274.

[6] I. M. Mitchell, "Comparing forward and backward reachability as tools for safety analysis," in *Hybrid Systems: Computation and Control*, A. Bemporad, A. Bicchi, and G. Buttazzo, Eds., 2007, pp. 428–443.

[7] E. Asarin, T. Dang, and A. Girard, "Reachability analysis of nonlinear systems using conservative approximation," in *Hybrid Systems: Computation and Control*, O. Maler and A. Pnueli, Eds., 2003, pp. 20–35.

[8] A. Girard, "Reachability of uncertain linear systems using zonotopes," in *Hybrid Systems: Computation and Control*, M. Morari and L. Thiele, Eds., 2005, pp. 291–305.

[9] M. Koschi and M. Althoff, "SPOT: A tool for set-based prediction of traffic participants," in *2017 IEEE Intelligent Vehicles Symposium (IV)*, June 2017, pp. 1686–1693.

[10] M. Althoff, "Reachability analysis and its application to the safety assessment of autonomous cars," Dissertation, Technische Universität München, München, 2010.

[11] B. Kim *et al.*, "Probabilistic vehicle trajectory prediction over occupancy grid map via recurrent neural network," *CoRR*, vol. abs/1704.07049, 2017.

[12] F. Giovannini, G. Savino, and M. Pierini, "Influence of the minimum swerving distance on the development of powered two wheeler active braking," in *22nd ESV Conference*, June 2011, paper 11-0258.

[13] C. Ackermann, J. Bechtloff, and R. Isermann, "Collision avoidance with combined braking and steering," in *6th International Munich Chassis Symposium 2015*, P. Pfeffer, Ed., 2015, pp. 199–213.

[14] C. Choi and Y. Kang, "Simultaneous braking and steering control method based on nonlinear model predictive control for emergency driving support," *International Journal of Control, Automation and Systems*, vol. 15, no. 1, pp. 345–353, Feb 2017.

[15] H. B. Pacejka, "Chapter 1 - Tire characteristics and vehicle handling and stability," in *Tire and Vehicle Dynamics (Third edition)*, H. B. Pacejka, Ed., 2012, p. 5.

[16] R. Schubert, E. Richter, and G. Wanielik, "Comparison and evaluation of advanced motion models for vehicle tracking," in *11th International Conference on Information Fusion*, June 2008, pp. 1–6.

[17] S. Söntges and M. Althoff, "Computing the drivable area of autonomous road vehicles in dynamic road scenes," *IEEE Transactions on Intelligent Transportation Systems*, vol. 19, no. 6, pp. 1855–1866, June 2018.

Operation in Tunnels Construction Works with Autonomous or Tele-operated Trucks

Felipe Jimenez, Jose E. Naranjo
University Institute for Automobile Research (INSIA-UPM)
Madrid, Spain
e-mail: felipe.jimenez@upm.es

Miguel Martín, Antonio Ramírez
SACYR Construccion
Madrid, Spain

Miguel Anguera, Pablo García
CAVOSA
Madrid, Spain

Abstract— **It is possible to transfer the technological base of autonomous vehicles to other areas. Such is the case of public works and, specifically, tunnels given their particularities. In this regard, this area is characterized by the fact that work is developed in a limited area, in which vehicles should not be registered for using them in public roads and the presence of human beings is restricted or prohibited. In this context, the implementation of autonomous vehicles in public works can provide significant improvements. For this, it is intended to improve the process of material removal from the tunnel by means of an automation and communication kit for heavy public works trucks. Then, the conventional vehicles can be transformed to work autonomously in a coordinated way between them and with other vehicles that operate also autonomously or manually.**

Keywords-autonomous vehicle; public works truck; V2X communications; teleopration; autonomous guidance.

I. Introduction

Almost all major car manufacturers, as well as other players outside this industry, are involved in the development of vehicles with a certain degree of autonomy, in addition to driver assistance systems, which, in some cases, take advantage of actuators automation. The results suggest that the automotive industry has opted for this type of technology for the not too distant future, although there is still a long way to go [1].

Autonomous driving is not confined to highways. There are several areas where the use of this technology can be applied, and special applications based on this type of vehicles have been implemented over the years. In this way, there are specific applications for off road environments, military missions, rescue, supervision and surveillance, land exploration, agriculture, etc. In general, all these applications share the fact that some tasks should not be made by a human operator due to the exposure to imminent danger or automation could provide a specific service with better benefits.

This paper presents the implementation of the elements for transforming a set of conventional public works trucks for tunnels construction into a set of autonomous vehicles that could operate both automatically or teleoperated from a control site. The main system architecture is shown, and preliminary results are discussed.

Section II includes a review of previous works on this specific field. Section III presents the specifications required for the system operation. Then Section IV shows the technological solution used for each of the elements that involve the system. Finally, Conclusions and current state of the project and future works are commented in Section V.

II. Autonomous Vehicles for Construction Purposes

In addition to public roads transport, autonomous driving has applications in specific structured scenarios looking for a reduction of the presence of the operator. Now we review some of these areas, such as mining and construction [2].

Mining jobs are highly demanding and workers' conditions can be extreme. In this sense, automation has a clear positive impact and has already been undertaken for a long time, to reduce operating costs and risks for operators. For example, fully autonomous mining trucks complete a set of tasks without the intervention of the operator and, instead, are monitored in a remote-control site by the miners to ensure that the trucks are operating efficiently throughout the mine. This solution increases productivity.

Among other examples, we can mention Sandvik, which has been developing loaders and autonomous trucks for mines over the past 20 years. Sandvik has developed vehicles that can automate the entire production cycle. Sandvik tests its vehicles to ensure they are safe and functional in an underground mine in Finland and also works closely with its customers to ensure that product expectations are met. Also, Autonomous Solutions, Inc. offers autonomous solutions for trucks, excavators, etc. Cyngn is another company that proposes autonomous excavators and loaders. Caterpillar is committed to assisting operator technologies that control specific functions of the machine to increase productivity and reduce costs, remote control systems that keep operators away from the cabins and even, in some cases, by totally autonomous trucks. Another example is the use of remotely supervised trucks at the Pilbara iron mines in Australia. 80 Komatsu trucks are used and each one has worked more than 700 additional hours compared to a conventional driven vehicle.

The negative aspects are the fact that the configuration, operation and maintenance of equipment without a driver is expensive. In addition to the vehicles themselves, there is the additional cost of mapping the mines to operate the vehicles, as well as the installation of a control center.

In the construction field, using robotic systems is a reasonable option in the short term since construction sites are closed to traffic and people. In addition, construction robots can work independently once guidelines are given. One of the main advantages lies in the fact that they offer safer jobs and less exposed to dust or vibration conditions, by managing the operation remotely. Thus, machines can autonomously perform dangerous and repetitive tasks.

However, complex and changing working conditions can reduce the potential benefit, which, together with the inertia of the sector, means that the presence of autonomous tools is still small, but not non-existent.

In October 2017, Built Robotics developed a loader controlled from an iPad that provides the functionality of entering operating parameters and, through satellite positioning and other sensors, it is able to perform the work. In the same year, Volvo introduced the HX2 electric charging vehicle and the LX1 hybrid loader, which provide considerable savings in energy consumption. Volvo also incorporates autonomous capabilities in the L120 excavator and the A25 articulated truck. In the same line, Caterpillar 793F trucks provided an increase of 20% of productivity in comparison with the one driven manually. Komatsu offers the semi-autonomous D61i-23 dozer, while Caterpillar and John Deere work in similar vehicles, which will start being semi-autonomous to evolve towards full automation.

III. System Specifications

The project presented in this paper corresponds to the technological line of transferring the fundamental knowledge acquired for road environments to scenarios in the industrial sector. This project aims to apply the knowledge of vehicle automation, positioning, obstacle detection and communications to a tunnel excavation, more specifically, in the tasks of extracting material from a tunnel to the area of intermediate discharge near its exit. This operation has a series of safety constraints that limit the human presence, and, on the other hand, the operation is very well defined, so it is susceptible for automation. As main challenges, beyond the type of vehicle, quite different from road vehicles, we could highlight the environment detection (considering light and dust), positioning (because of the lack of Global Positioning System (GPS) signal) and collisions avoidance, as well as the automatic and remote management of the points of material loading and unloading, and the synchronized operation of a small group of vehicles working simultaneously.

The project aims to replace conventional vehicles driven manually by autonomous and connected vehicles, also managed from a control center. Specifically, the aim is to automate 3 Volvo A-25 public works trucks (Figure 1), as well as including communication systems and a user interface to guarantee compatibility with other manually driven trucks, so that they can share the workspace and the operation in a coordinated way from a monitoring and control center. The 6x6 Volvo A25 has great capacity and flexibility in mixed and very difficult terrain conditions.

Figure 1. Volvo A-25 public works truck

Excavation of a tunnel is usually executed by means of different machinery such as a front wheel loader, model Volvo L120 or similar. These machines load the material on articulated trucks such as Volvo model A25D. Once arrived at the end of the tunnel or near crossing zones enabled for it, these trucks will turn around so that their cabin is oriented facing the exit of the tunnel. Then, in reverse gear, they will be properly positioned so that the loading machines can pour the material into the truck's box. Once the load is completed, the truck will move to the corresponding dumping area outside the tunnel. The rest of the trucks of the fleet will be returning and/or waiting inside the tunnel, in the areas enabled for it.

This process can be automatized because of its repeatability. To indicate to the trucks the final stopping points, both inside and outside the tunnel, beacons are placed, which allow the operators to easily modify the collection and unloading points.

We can define 5 specific innovations of the project derived from the development of this project:

- Development of a kit for automation of Volvo A-25 trucks, which will equip them with the ability to operate autonomously. The vehicle will maintain its ability to be driven manually and, in addition, will allow autonomous driving. Furthermore, since automation is provided by a removable kit, it can be installed in conventional machinery that is currently working on the construction site so specific machinery is not required.
- Development of a perception system for operation in off-road environments in tunnel construction without positioning information, and under poor light and dust conditions.
- Development of cooperative capacities, so a set of vehicles can work in the same area at the same time in a coordinated manner. To do this, automation will be carried out in three vehicles and communication systems will be incorporated to enable them to exchange information in real time with each other and with the management and monitoring system.
- Development of a management system for monitoring the fleet of autonomous vehicles during

the operation in the excavation of a tunnel in order to control in real time the development of the same.
- Development of a methodology for the automation of public works vehicles so that the results of this project can be extended to other machinery and other machinery manufacturers.

IV. Technological Solution

The technological solution involves 5 main elements:
- Automation kit for transforming a conventional vehicle into a vehicle with autonomous or teleoperated capabilities (Subsections A and B).
- Low-layer control system. This element involves the steering system automation, speed automation and control subsystem, considering all the information and commands provided by sensors and communications (Subsection C).
- Perception and High-level control system. This system includes the information collection (vehicle positioning, obstacles detection and reference element detection) and processing, as well as the guidance system (Subsection D).
- Teleoperator site from which vehicles are supervised and controlled (Subsection E).
- Communications systems for exchanging information between vehicles, and with the control site in order to achieve a coordinated operation when more than one vehicle is involved (Subsection F).

Figure 2 shows a block diagram with the interconnection of these elements, including a new emergency system to stop the vehicle externally in case of failure or incorrect operation.

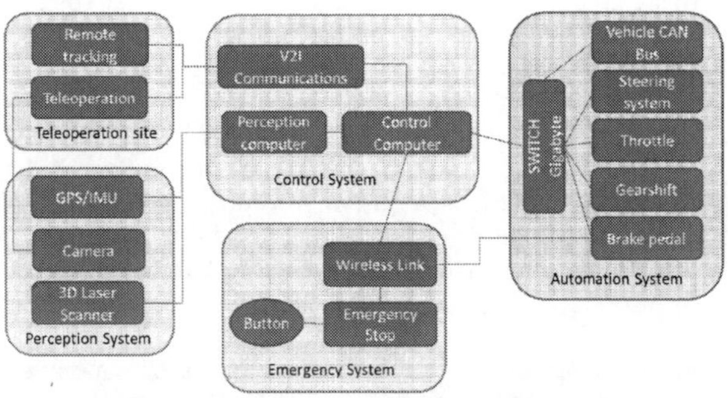

Figure 2. Device for controlling the steering system

A. *Steering automation*

The Volvo A25 truck has articulated steering, with an exclusive Volvo steering system that is self-compensating hydromechanically and with re-coupling between the rear axle and the steering slide from the steering and tilt valve. The steering system is coupled in parallel with the tilt system, with which it has common hydraulic pumps.

There are two ways to act automatically on the steering system: 1) act directly on the hydraulic cylinders, inserting a bypass on the current hydraulic circuitry with solenoid valves controlled from the computer; 2) attach an actuator on the steering column that is capable of moving the steering wheel, not modifying the pre-existing equipment of the truck.

The second one is chosen. To do this, a motor must be attached to the steering column in such a way that it exerts the same effect that human drivers do when moving the steering wheel. Additionally, an absolute encoder must be added to the steering bar indicating the angle that the steering of the vehicle is turned, in order to be able to close the control loop.

The solution adopted is based on the patent ES 2516568 B2, "Equipment to automatically control the steering system of a vehicle", property of the Technical University of Madrid [3] (Figure 3). It includes the following features [4]:
- The device is independent of the type of vehicle and the type of steering assistance system (electric or hydraulic), including the presence or absence of such assistance, unlike other previous developments that require a specific type.
- The assembly of the device does not imply any permanent modification of the vehicle, nor does it eliminate the steering column.
- Orders through the vehicle internal communications bus are not required for the vehicle control, which allows the use of the device in every type of vehicle.
- The coupling or decoupling of the autonomous mode is controlled automatically and not manually, so it could be carried out while the vehicle is moving and instantaneously, using the same microprocessor that controls the rotation angle of the steering wheel.
- The driver does not lose control of the vehicle.
- It is not visible and does not interfere with the driving task.
- The device is removable.

Figure 3. Device for controlling the steering system

With the solution used, in normal driving, the driver acts on the steering wheel, the electric motor is stopped and an electromagnet, which acts as a clutch, is deactivated so no effort is transmitted. When it is decided to change to autonomous driving, the electromagnet is activated, and the rotation of the electric motor is transmitted to the steering column. This connection can be made at any time and with any position of the steering wheel. On the other hand, if the vehicle is in autonomous mode and the driver wishes to perform different actions than the ones intended by the system, the system could be deactivated as soon as the driver acts. The device performance has been tested on other vehicles types [4] with satisfactory results. The electric motor, gears and mechanical parts have been designed in this case considering the measured forces involved.

B. Speed automation

The throttle of the Volvo A25 is electronic. In this case, in order to automate it, it is necessary to transmit the analog signal proportional to the angle of the accelerator pedal from an external source. Then, it is necessary to send an alternative analog signal generated by a computer that emulates the one generated by the original potentiometer and allows its operation from the automation system.

On the other hand, the Volvo A25 is equipped with a classic brake assist system with hydraulic assistance. This implies that access to related electronics is not enough for its automation, so it is necessary to act mechanically on some of its components in order to obtain the desired action.

Finally, the Volvo A25 equips an automatic gearbox with a torque converter, where the driver can select from 6 different positions. The action of the gear selector on the Electronic Control Unit (ECU) is totally electronic, activating each selection according to a series of 8 bits that correspond to the electronic outputs of the selector itself. In order to proceed with the automation of this element, it is necessary to bypass the outputs of the gear selector in order to emulate them from an electronic digital output card.

C. Control architecture (low-level control layer)

The control scheme must define the necessary equipment requirements to carry out the automation of the vehicle's actuators: throttle, brake, steering wheel and gearshift, as well as their interconnection and operation from a centralized on-board computer.

In this way, the control architecture has been designed in a distributed manner, where each actuator constitutes a subsystem and all components are interconnected in two levels. On the one hand, all the components of each subsystem will be connected through a real-time Controller Area Network (CAN) bus line. This architecture simplifies the electronic layout and centralizes all the commands in the same protocol for all the actuators. In case any modification would be introduced in the future, it is much simpler and modular. On the other hand, each subsystem will be interconnected and connected to the central computer by a TCP/IP network. Then, the malfunctioning of one of the subsystems will not cause the global system to fail, since the architecture makes it tolerant to failures in its components.

In this way, we define 3 subsystems:

- Subsystem of access to vehicle information. Access to the data of the internal bus of the vehicle, in order to obtain the data of vehicle speed, engine speed and change of gears. An Ethernet/ CAN card, connected to the local network of the vehicle, is used for this purpose.
- Subsystem of actuators driven by motors. In this case, this subsystem controls the steering wheel and the brake, two elements with the same configuration: they are driven by DC motors mechanically connected to the controls. These motors will receive orders directly through a CAN line through CANOPEN protocols. An Ethernet/CAN card is used for this, connected to a second ad-hoc bus, which will not interfere with the original CAN bus of the Volvo A25.
- Electronically controlled actuator subsystem. In this case, the throttle and the gearshift of the vehicle will be controlled by means of this subsystem, which obeys the analog electronic signal protocol. To do this, these signals will be generated by an analog output card, which is connected to the Ethernet network through another Ethernet/Serial card.

All subsystems are linked by a Gigabit switch, which ensures the maintenance of real time in the global system. In case this switch fails, the system detects that no more new data is received so a stopping command is sent automatically to the brake pedal, whose actuator receives power directly from the vehicle battery. This control system is implemented in a box containing all the electronic devices, bus interfaces and communication antennas inputs (Figure 4).

Figure 4. Electronic and communication control box

D. Perception and autonomous guidance system (high-level control layer)

The fundamental objective of the autonomous guidance system is to take control of the different actuators of the vehicle and provide them the appropriate commands to perform the desired tasks. In order to achieve this objective, the vehicle equips a GPS receiver for using this signal when available and installs 2 different perception systems:

- a 3D laser scanner sensor
- a camera placed near the human driver point of view

As GPS signal is quite poor or unavailable in tunnels, positioning and autonomous guidance is performed be means of the laser scanner. So, this sensor is used for improving positioning accuracy detecting the walls of the tunnel in order to follow the path safely. It is also used for obstacles detection [5]-[8] to stop the vehicle in case the path is blocked and send a warning to the teleoperator. The camera sends images to the teleoperation site in order to take control decisions. Then, this high-level control layer can work in two operating modes:

- Autonomous mode: a trajectory or a reference element is followed without human actions. For this purpose, previous algorithms as presented in [9] are used to improve accuracy and robustness.
- Remote control: commands are sent from the teleoperation site and they are translated to be transferred to the low-level control layer. In both cases, obstacles detection is active and impose its

decisions on the other commands because of safety reasons.

E. Teleoperation site

The tele-operator site is responsible for monitoring the trajectory and operation of all vehicles that circulate in the construction area [10]. It includes devices for controlling trajectory and speed of the vehicle in the tele-operated mode. Furthermore, one of the screens shows the images from the cameras placed on the vehicles near the driver position and the other screen presents operation data of the vehicles. Finally, in case the autonomous mode is activated, the mission tasks are stored in the control computer and sent to the vehicles. These missions could be updated in an easy way by the operator.

F. V2X communications

Communications allow the exchange of information between the teleoperated or autonomous vehicles and the teleoperator site. For this purpose, Vehicle-to-X (V2X) standard communications systems are used. The requirements that these communication systems must comply with are the following ones.

- Desired range: 1 km with direct line of sight
- Bandwidth of at least 1 Mbps
- Operation in broadcast mode.
- UDP/IP communications protocol in order to guarantee the robustness of the operations.

In this way, INSIA-ITS communications modules are used, which comply with the current standards to support communications in road vehicle environments and can be used with any type of IP protocol. In order to extend the range of these modules, they are equipped with signal amplifiers. These modules have been satisfactory tested in road scenarios [11][12] and trials in tunnels are expected for the near future.

V. CONCLUSIONS

This project aims to develop a cooperative work between autonomous trucks in a complex environment such as tunnel construction in order to limit human intervention in a highly aggressive environment. In addition, it has the particularity that autonomous vehicles are obtained transforming conventional manual driving with non-permanent nor intrusive adaptations, thereby respecting the possibility of dual operation, which represents a clear competitive advantage.

The project is a challenge for autonomous driving given the environment in which the work of the vehicles must be developed, which negatively affects perception and positioning, as well as having a very restricted space. On the other hand, communications between the vehicles and with the infrastructure and a control site allow the coordinated management of the vehicles to achieve greater productivity and anticipate potential dangers.

At the current stage, subsystems have been tested independently. In this sense, vehicle automation lower control layer has been tested, as well as the internal communication bus that provides control commands. Furthermore, the teleoperation site for controlling and supervising the vehicle operation is finished and communication with the vehicle has been tested. Perception algorithms are now under modification in order to adapt them to the specific scenario that can be found in a tunnel. As a final stage, coordinated operation between different trucks managed from the same teleoperation site is expected to be implemented.

ACKNOWLEDGMENT

This project has been partially financed by the Spanish Ministry of Science, Innovation and Universities (project "Tunnel Autonomous Driving, TUNNELAD", with code RTC-2017-6382-4) and the Spanish Ministry of Economy and Competitiveness (CAV project, with code TRA2016-78886-C3-3-R).

REFERENCES

[1] F. Jimenez.. "Retos tecnológicos en el desarrollo e implantación del vehículo autónomo y conectado". Carreteras, vol. 216, pp. 8-16, 2017.

[2] F. Jiménez. "Aplicaciones especiales de la conducción autónoma" Revista de Obras Públicas, in press

[3] F. Jimenez, J. E. Naranjo, M. González and O. Gomez. "Equipo para controlar automáticamente la dirección de un vehículo" Patent ES2516568, 2016

[4] F. Jimenez, J. E. Naranjo and O. Gomez. "Dispositivo universal para el control automático de la dirección de un vehículo". Dyna Ingeniería e Industria. Vol 89, 4, pp 398-404, 2014.

[5] K. Chu, M. Lee, and M. Sunwoo, "Local path planning for off-road autonomous driving with avoidance of static obstacles," IEEE Trans. Intell. Transp. Syst., vol. 13, pp. 1599–1616, 2012.

[6] C. Caraffi, S. Cattani, and P. Grisleri, "Off-road path and obstacle detection using decision networks and stereo vision," IEEE Trans. Intell. Transp. Syst., vol. 8, pp. 607–618, 2007.

[7] J. Larson and M. Trivedi, "Lidar based off-road negative obstacle detection and analysis," Proc. IEEE Conf. Intell. Transp. Syst. Proceedings, ITSC, pp. 192–197, 2011.

[8] E. Shang, X. An, T. Wu, T. Hu, Q. Yuan, and H. He, "LiDAR Based Negative Obstacle Detection for Field Autonomous Land Vehicles," J. F. Robot., vol. 33, pp. 591–617, Aug. 2016.

[9] F. Jimenez, M. Clavijo, J. E. Naranjo and O. Gómez. "Improving the lane reference detection for autonomous road vehicle control". Journal of Sensors. ID 9497524, pp. 1-13, 2016

[10] J.E. Naranjo, F. Jimenez, M. Anguita and J. L. Rivera. "Automation Kit for Dual-Mode Military Unmanned Ground Vehicle for Surveillance Missions". IEEE Intelligent Transportation Systems Magazine, in press.

[11] J.J. Anaya, E. Talavera, F. Jiménez, F Serradilla and J. E. Naranjo. "Vehicle to Vehicle GeoNetworking using Wireless Sensor Networks". Ad Hoc Networks, vol. 27, pp. 133 – 146, 2015.

[12] E. Talavera, J. J. Anaya, O. Gómez, F. Jiménez and J. E. Naranjo, "Performance comparison of Geobroadcast strategies for winding roads".Electronics, vol 7, pp. 1-15. 2018

A Survey of Autonomous Vehicle Technology and Security

Mustafa Saed
HATCI Electronic Systems Development
Superior Township, Michigan, USA
msaed@hatci.com

Kevin Daimi and Samar Bayan
University of Detroit Mercy
Detroit, Michigan, USA
{daimikj, bayansa}@udmercy.edu

Abstract—Fully autonomous vehicle will soon be a reality. This will present a vector of issues and challenges including economic, social, safety, environmental, and security problems. Security will participate in enhancing safety of passengers and pedestrians. With the current non-autonomous vehicles, the work on security is ongoing and mainly in its research status. The sophisticated technology of autonomous vehicle will furnish a path of even more complicated security issues. In this paper, the autonomous vehicle technology and its security will be briefly surveyed to allow researcher the opportunity for further research in this field. In particular, the paper will address the five levels of autonomous vehicle and its current state with regards to these levels, the internal architecture of the vehicle, and security threats facing this vehicle technology.

Keywords—Autonomous Vehicle; Autonomous Vehicle Architecture; Level of Autonomy; Security Attacks; Security Defenses.

I. INTRODUCTION

Vehicles were first invented to facilitate the transportation of people. In 1769, the first stream road engine was invented by Nicolas-Joseph Cugnot [1]. A few years later, vehicles used the internal combustion engines powered by hydrogen and oxygen mixture [2]. Vehicles became gasoline powered in 1885 [3], and the first true electric car was invented in 1888 [4]. Electric cars were popular between the 19th and 20th centuries due to their level of comfort and ease of operation. Since then, vehicles had been improved so much especially with the introduction of artificial intelligence in 1960 [5]. Researchers started to think of ways to overcome the driver's role, so they added autonomy to vehicles. Autonomy allowed vehicles to be categorized from a regular vehicle with no autonomous characteristics to full autonomous vehicle capable of moving by itself. In 2019, level 3 autonomous vehicle, Tesla Model 3, had been introduced to the market [6].

Autonomous vehicles include the classical vehicle characteristics with the additional autonomy flavor. They are expected to collect enormous data from various sources and replace humans in driving. These accumulating data will be huge and will open further research venues for many fields including technological, data science, and security. With full autonomy, humans are no longer needed to control the vehicle's movements. However, autonomy as defined by National Highway Traffic Safety Administration varies depending on the way the control functions are handled by the vehicle. The full autonomous vehicle extracts information from the surrounding environment via various signals, analyzes these signals and executes appropriate path of movement [7]. This implies that in all the phases of this procedure, human will not even play any role in the environmental perception. With these high control functions, the vehicle becomes more dependent on communication networks internally and with exterior environment [8]. This exceedingly reliance on communication networks will unlock the gates for even more sophisticated security attacks. There are two types of communications in autonomous vehicles, Inter-vehicle and intra-vehicle communications [8]. Intra-vehicle communications, represented by buses, are responsible for data transfer between the autonomous vehicle's components. Inter-vehicle communications deals with transferring of data between the vehicle and the external environment including other vehicles, infrastructure and smart road signs. This makes the autonomous vehicle more vulnerable to various security attacks that are classified based on type of the attacker, motivation for the attack, type of the attack, and the target for the attack [9]. Consequently, the attacker will be able to collect information from the autonomous vehicle, modify it, and cause harm for both vehicles, their passengers, and possibly passengers of other vehicles. Thus, innovative and leading-edge security measures will be demanding due to the sophistication of the communication process.

To ensure autonomous vehicle network security and avoid potential attacks, different defenses have been proposed. These security techniques satisfy a collection of requirements pointed out by data integrity, data confidentiality, user and in-vehicle authentication, and availability [10]. For this reason, new cryptographic techniques should be established to enhance the autonomous vehicle's security and ensure that the original data is not altered to make certain vehicle's performance will not deteriorate and the safety for all is granted.

This paper deals with surveying the current and future technology of autonomous vehicle and its security. To this end, the levels of autonomous vehicles are introduced in Section II. Section III presents the architectural technology of autonomous vehicles, and the threats that autonomous vehicles are vulnerable to are explained in Section IV. Autonomous vehicle security is covered in Section V. The paper is then concluded in Section VI.

II. AUTONOMPUS VEHICLE LEVELS

The mission of full autonomous vehicle is to transport passengers to their destination without the need for a human driver. The National Highway Traffic Safety Administration (NHTSA) defines autonomous vehicle as "those in which at least one aspect of safety-critical control function occurs without direct driver input" [11]. This definition reveals that autonomous vehicles are categorized by levels ranging from Level 1 to Level 5 [12] [13]. According to NHTSA, Cruise control, automatic braking, and lane keeping are considered examples of automation systems, or safety-critical control functions. The National Highway Traffic Safety Administration does not consider vehicles equipped with vehicle-to-vehicle services for safety warnings as autonomous vehicle. Level 0 refers to vehicles with no autonomy. The driver in Level 0 autonomous vehicles has full control over all tasks within the vehicle. Both NHTSA and Society of Automotive Engineers (SAE) [11] [14] depicted the levels of autonomy as shown in Figure 1.

The five levels of autonomous vehicle represent the various magnitude of automation that the vehicle is equipped with. The transition from a lower level to a higher level signifies the increase in automation. This style will continue until full automation (Level 5) is reached.

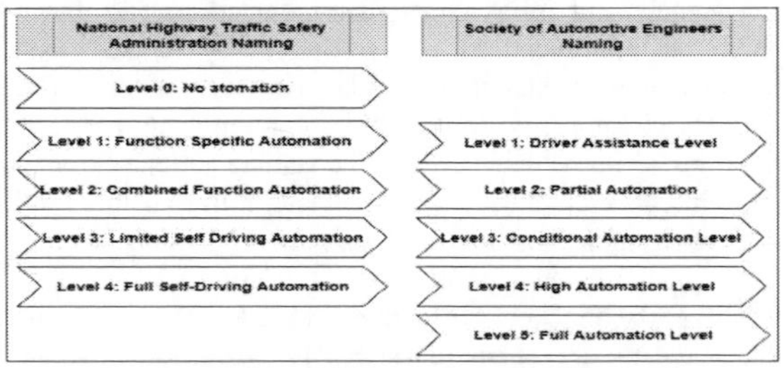

Figure 1. NHTSA and SAE Autonomous Vehicles Levels

A. Level 1 - Driver Assistant

At this level, the driver is responsible for monitoring the outer environment and taking decisions to control the vehicle's movements. The system shares only one of the control functions with the driver: steering and acceleration speed, or braking control [15]. Human will execute most of the driving tasks [16]. An example of Level 1 autonomous vehicle is Jaguar Land Rover. The vehicle is responsible for off-road cruise control. In an off-road cruise control, the driver is responsible for steering while the system will set the appropriate speed. The maximum speed is predetermined by the driver [17].

B. Level 2 - Partial Automation

The automated system is in charge of two of the primary control functions of driving at this level [11]. It fully controls the steering and acceleration tasks with limited driving conditions [18]. The driver handles the remaining tasks including environmental monitoring. An example of Level 2 autonomous vehicle is Tesla Model S [19]. Tesla introduced a technology that makes the vehicle capable of accelerating, maintaining lane position, and parking without the help of the driver. The driver is only responsible for holding the steering wheel and monitoring the environment [19].

C. Level 3 - Conditional Automation

The automated system is fully responsible for monitoring the environment and performing the safety-critical functions. The automated system will handle driving and monitor the envrionment. The autonomous vehicle anticipates the driver to accomplish backup for the system and resume driving when needed [1]. The difference between Level 2 and Level 3 autonomous vehicles is that at Level 3 the driver may not constantly monitor the environment during driving. The system can share this task with the driver [11].

D. Level 4 - High Automation

The vehicle at this level is fully charged to control driving. The difference between Level 3 and 4 is characterized by the needed interference of the driver in case of failure. This implies that Level 3 system expects the intervention of driver for backup, but Level 4 system works without any expectation from the driver [12]. This level has some limitations determined by maximum speed, and low speed, and adverse weather conditions, such as snow falling [20].

E. Level 5 - Full Automation

Level 5 does not expect the vehicle to have steering wheel and performs all environmental analysis and planning techniques to reach destination by itself [18]. Level 5 vehicles are similar to level 4 but with with no limitations [20]. The vehicle at this level will no longer need steering wheel, pedals or human to control tasks [16]. Google is working on building Level 5 autonomous vehicle through its company Waymo [21]. This level could be referred to as full vehicle automation.

III. Current State of Vehicle Autonomy

Autonomous levels describe the role human plays while driving. However, Level 4 and 5 autonomous vehicles are not implemented yet. This is due to the difficulty of making the system totally reliable on itself without expecting human interference. This means that the vehicle will fully analyse and even take care of any failure case [22]. At the present time, the autonomous vehicle technology lends itself to Level 3 autonomous vehicle. In Table I, autonomous vehicles currently manufactured by auto industry are demonstrated. Moreover, many manufacturers announced that they will have Level 4 autonomous vehicles available in year 2020-2021 including Toyota, Volvo, Renault-Nissan, Hyundai, and Ford [24].

TABLE I. RECENT AUTONOMOUS VEHICLES AND THEIR LEVELS

Manufacturer	Mobileye [23]	Tesla [24]	Audi [25]
Model		Model S	A8
Automation Level	Level 2	Level 2	Level 3

IV. Autonomous Vehicle Threats

Understanding the autonomous vehicle threats stems from understanding the sophisticated autonomous vehicle technology and architecture. The autonomous vehicle needs to analyze data from the surrounding environment. These data are collected from perception sensors, other vehicles and various smart infrastructures [7].

A. Autonomous Vehicle Architecture

When analyzing the security of a system and identifying the associated threats, it is essential to understand the underlying architecture to establish the needed security protocols. The way autonomous vehicle analyses things is similar to people's action-perception technique. The approach consists of perception, planning and control systems [7] [8] [26]. First, the perception system is responsible for sensing the environment and finding out the location of the autonomous vehicle [27]. The location can be represented in three ways; relative location, absolute location, and hybrid location [7]. Relative location is calculated by adding the distance and orientation of the vehicle to the initial position. The global positioning system, GPS, is in charge of providing the absolute vehicle location. Hybrid location is a mixture of both, relative and absolute locations. The goal is to find the real-time efficient location. Autonomous Vehicle uses the hybrid location technique to localize itself [7]. Sensing the environment is represented by lane line identification, obstacle detection, and road signs analysis. This is delivered through cameras, LIDAR, and Radar [8]. The Light Detection and Ranging (LiDAR) supplies high-resolution, three-dimensional information about the vehicle's surrounding environment. Having completed the perception, the Planning System picks up data from perception system, analyzes it and makes the appropriate decision for movement. The input for this subsystem is a combination of the perception system's output data, feedback from the control system, and the Inter-vehicle communication data. Finally, the Control System implements the decision taken by the Planning System through a large number of Electronic Control Units (ECUs). This PPC architecture (perception, planning, and control architecture) is similar to perception, cognition, and action systems of the humans [28]. Details of these systems and their relationships are depicted in Figure 2.

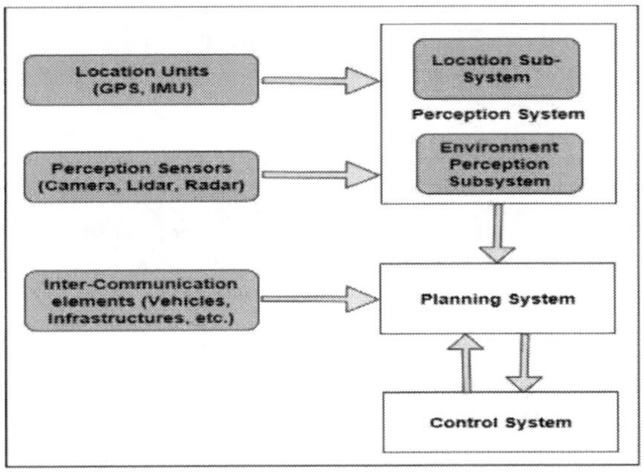

Figure 2. Internal Architecture of Autonomous Vehicle

For proper performance, each system needs three components; sensors, processors and communication technologies [29]. The autonomous vehicle, as shown in Figure 2 above, collects data from onboard sensors, such as camera, lidar and radar, and from outer components including vehicles and infrastructure. The communication technologies within these two categories are referred to as intra-vehicle communication and Inter-vehicle communications. Inter-vehicle communication allows vehicle's parts to communicate and exchange information. It employs different buses to achieve this communication, such as Control Area Network (CAN), FlexRay, Local Interconnect Network (LIN), Media Oriented System Transport (MOST), and ethernet [8]. The bus technologies used by two selected Autonomous Vehicles (AVs) are shown in Table II below.

TABLE II. AV INTRA-VEHICLE COMMUNICATIONS BUSES

	Tesla Model S	Audi A8
Level	2	3
Technologies	CAN, LIN, Ethernet [8]	CAN, LIN, FlexRay, MOST [8]

Furthermore, the autonomous vehicles can collect real-time data from everything around it to enhance decisions taken by its planning system. These relations are categorized as vehicle to vehicle (V2V), vehicle to infrastructure (V2I), vehicle to road signs (V2RS), vehicle to internet of things (V2IOT) and vehicle to everything (V2X). Inter-vehicle networks are divided into low power technologies, such as Bluetooth and Zigbee, and IEEE 802.11 family technologies including WiFi and Dedicated Short Range Communication (DSRC), and base station driven technologies, such as Worldwide Interoperability for Microwave Access (WiMAX) and LongTerm Evolution for Vehicle (LTE-V) [8]. As depicted in Figure 3, the most popular networks, LTE-V and DSRC, currently deployed by the autonomous vehicles are demonstrated.

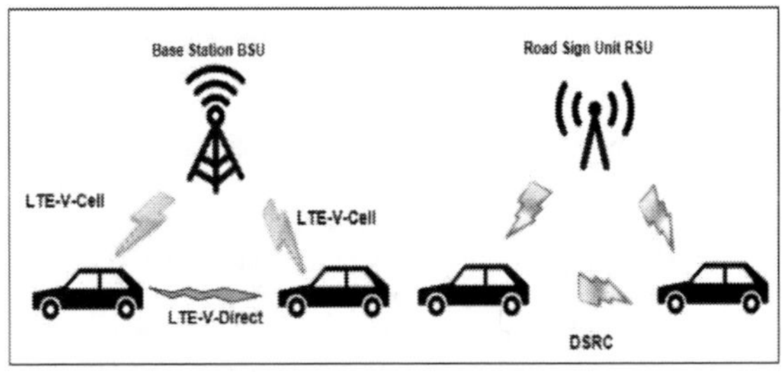

Figure 3. Inter-Vehicle Communications- LTE-V and DSRC

Hence, the autonomous vehicle is vulnerable to security attacks due to increased internal communication (intra-vehicle communications) and Inter-vehicle communications.

B. *Autonomous Vehicle Threats*

The autonomous vehicles are coupled to different communication techniques. This makes it vulnerable to different types of attacks. For example, the vehicle-to vehicle V2V connectivity in the autonomous vehicles increases the ability of the attacker to join multiple vehicles in a vehicular botnet [30]. The vehicular botnet is a collection of networked vehicles controlled by the attacker [31]. Here multiple bots will join botnet and will execute whatever the attacker instructs them to do to the vehicle network. Various attacks are provided in Section V.

Other types of security attacks include password and key attacks, Denial of Service (DoS), Network protocol attacks, and Rogue Updates attack [32]. The first one is classified into dictionary, rainbow and brute force attacks. The attacker tries multiple attempts by using list of words, precomputed hashes or alpha-numeric combination to crack the password [32]. In other words, the attackers try to discover passwords by using every possible password commonly used and stored in a predefined file or database. In brute force attack, attackers try every possible combination of letters, digits, and other characters to discover the password. Finally, Rainbow attack consumes less time than the other two types by using a large store of precomputed hashes and comparing the stolen hashed password with those in the store.

In autonomous vehicle, DoS attack can be achieved through single node (within a vehicle), V2V communication or V2I communication [33]. A bogus Electronic Control Unit (ECU), which can be any device that succeeds in communicating with the ECUs, can send huge number of messages to other ECUs (single node), and a vehicle or group of vehicles can initiate large number of messages to a vehicle or even the infrastructure (V2V or V2I). Even worse, a bogus infrastructure can bombard vehicle with many messages (I2V). The Dedicated Short-Range Communication (DRC) and Wireless Access in Vehicular Environments (WAVE) are used as communication mediums that enable messages exchanged between vehicles and the entire Vehicle to Infrastructure (V2I) environment. Several types of attacks associated with wireless V2X communication protocols have been demonstrated by security experts that could disrupt the availability and the performance of autonomous vehicles. Potential Attack Scenarios include an attacker jamming the main wireless communication medium so that the network will no longer be available for legitimate users. This would cause a DoS attack that prevents authentic users (autonomous vehicles) from being able to communicate with each other or with the whole infrastructure. Furthermore, an attacker can achieve a DoS attack by generating a high volume of false messages which flood the network impeding the performance of any decision-making processes of the autonomous vehicle. An attacker can compromise the communication network and change the content of a warning message or send fake messages to other vehicles to disrupt the smooth road functionality or cause accidents. For example, an attacker who receives a warning message "Road Constructions Warning" from a nearby vehicle can change the content of the message and send the message "Road is clear" instead. Also, an attacker can compromise the confidentiality of autonomous vehicles' operations by eavesdropping. An example of this attack could be the collection of location and routing information of specific autonomous vehicles in order to further harm the passengers in these vehicles or even steal the vehicles. Due to the fact that the exchanged messages are encrypted, this attack would require decrypting the exchanged messages to be successful.

The above scenarios lead to great damage to the autonomous vehicle communication system due to the bulk amount of data being sent. In network protocol attack, the attacker analyses the protocols to find out the weak points that can be exploited. Some researchers showed that CAN and FlexRay protocols are most vulnerable to this type of attacks. Rogue Updates attack occurs when the software of Electronic Control Units (ECU) are updated by versions not from the automakers (manufacturer). The Control Area Network (CAN) bus has limited number of bits (64 bits) dedicated to message transmission. This limitation does not allow strong encryption of messages and the authentication of these messages and their senders. Therefore, it is vulnerable to such attacks and other types of attacks as specified below. FlexRay protocol is based on both the Physical and Data Link layers. All the possible attacks on these two layers, including DoS, find their way to the FlexRay protocol. For this protocol, only one node can produce the main signal while other nodes can only create idle signals. DoS will be achieved by continuously sending the main signal.

Modern vehicles have over fifty attack points (over 50 generic attack points that hackers can exploit in order to attack a vehicle) including the in-vehicle systems (CAN, FlexRay, Ethernet communication protocol), the Mobile Network Operations (MNO) and the backend operations. Due to the additional technology that is introduced into autonomous vehicles, the number of attack points is increasing. At the same time, an increasing number of hacking tools, including software and hardware specifically designed to monitor and control the in-vehicle network, systems, and applications, are becoming available. Hacking tools can be used by researchers or hackers that are interested in exploiting the vehicle environment for their own benefits. Most of them are becoming open-source and are available for free. More importantly, the risk of a successful attack to an autonomous vehicle is dramatically increasing, since there is no fallback. Vehicle operations rely on technology and a potential compromise could have fatal consequences.

Attacks that result from increased connectivity of the autonomous vehicle are classified in to Physical Access Attack, Close-Proximity Attacks, and Remote Access Attacks.

Physical access attacks are categorized into Invasive and Noninvasive attacks [9] [34]. This classification is based on whether the attack is through device mounted on the autonomous vehicle or not. Invasive physical attacks are subdivided in turn into Side Channel Attacks, Clock Glitch Attacks, and Power Glitch Attacks [33]. Side Channel Attack occurs when the attacker builds up an alternative path for the data [9]. In Clock Glitch Attack, the signals from instruction sequence of the modules are inspected. The latter then is injected with fault signals resulting from exploiting timing violations [35]. Power Glitch Attack is done through analyzing the power consumptions of the electronic control units [36].

Invasive Physical attack exploits the weak points within the autonomous vehicle. Examples of such points are the Onboard Diagnostic Unit OBD and the media system [32]. All cars made after 1996 are required to have an Onboard Diagnostics Board connection (OBD-II) located within two feet of the steering wheel. All vehicles manufactured after 2008 must share the same OBD-II protocol. The OBD-II's initial function was to monitor mandated emissions equipment. Today, the port is used to monitor and control multiple functions. Service personnel plug equipment into this port for both diagnostics and ECU programming, typically via Windows-based computers, creating at least two paths for the introduction of malware. First, dealership computers typically connect to the Internet (often required by manufacturers) for daily code/firmware updates. During that process, malware could be downloaded and affect their computers. They in turn could spread the malware when they connect them to a vehicle's OBD-II port. A second path is accomplished through hacking into the dealership's wireless network. In addition to dealerships and mechanics, parents can connect an app to the OBD-II port to remotely monitor their children's driving, and fleet managers use apps to keep track on how their fleet vehicles are being driven. These are further sources of attacks through the OBD-II protocol. Not only hackers intend on introducing malware, but clever thieves can access the port to clone "smart keys" and simply drive away with a stolen car. Attackers exploit these vulnerabilities to have access to the internal communication buses. Both Onboard Diagnostic Unit and the media system are connected to the CAN bus in the autonomous vehicles [9]. These invasive attacks are classified into Code Modification Attacks, Code Injection Attack, Packet Sniffing, and In-vehicle Spoofing [32]. Code modification attacks are characterized by modifying the codes transferred through CAN bus. Code Injection Attack works in a similar way by injecting harmful codes through CAN bus. Packet Sniffing is a passive attack that allows viewing transmitted data between modules for the purpose of collecting information. In-vehicle Spoofing is also a passive attack where in which attacker is masquerading or pretending to be another identity to modify data [9]. These are executed through mounted devices and exploits the internal communications.

Remote Access Attacks are represented by the ability of the attacker to control the vehicle remotely. They evolve as a result the expansion of wireless communications in the autonomous vehicle in addition to the growth of external interfaces including the smart cameras and Lidar. They can be categorized as Malware Injection, Signal Spoofing, and Fault Injection Query Attacks [34]. Signal Spoofing attacks exploit the external communications. These include GPS spoofing in which the attacker broadcasts incorrect GPS data [9]. Malware Injection attacks can be considered as code injection attack through the external wireless communications. Here the data is injected through these connections. These can be successful through exploiting the external communications of the autonomous vehicle.

V. Autonomous Vehicle Security

Autonomous Vehicle Security can be ensured by using various strong encryption and authentication algorithms and techniques to minimize the attack surface. Designing security for a system follows a number of steps: determining the objective, assessing the sensitivity, estimating capabilities, and determining the control features [10]. Various security attacks on AV are illustrated in Figure 4. The security requirements for autonomous vehicle are as follows: authentication, data confidentiality, data integrity, authorization, privacy, and traceability (tracking the malicious entities) [10]. These are depicted in Figure 5.

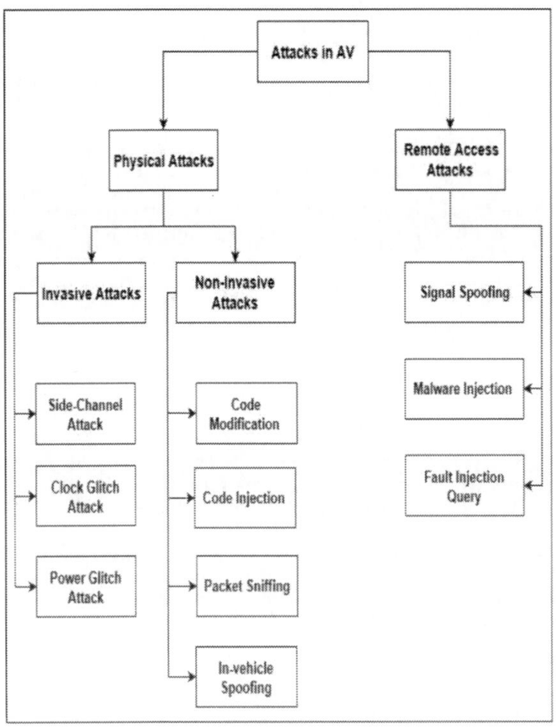

Figure 4. Security Attacks on AV

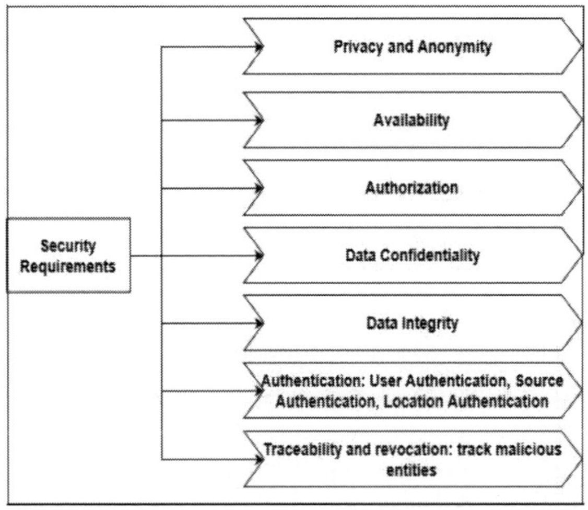

Figure 5. Security Requirements for Data Transfer in AV

Note that all the attacks mentioned in Figure 4 can impact all levels of autonomous vehicle. However, as the degree of automation increases (moving from lower level to higher level), the impact of these attacks becomes more severe.

Autonomous Vehicle (AV) defenses are classified into four categories; Active Defenses, Preventive Defenses, Passive Defenses and Collaborative Defenses [9] [34]. These are clarified in Figure 6. Preventive defense tends to stop the attack when it occurs by increasing the security measures. This type of defense includes authenticating the user and the in-vehicle device, securing the communications, and controlling the network traffic (through firewall). These are implemented to ensure data confidentiality through using symmetric and asymmetric encryption processes and enforcing message integrity through the use of Message Authentication Code (MAC) or hash techniques.

Active defenses can be done by continuously monitoring the security scales of the autonomous vehicle or by applying adaptive security. The latter is characterized by reconfiguring the attack targets and improving tactics to have better control when the attack occurs [9].

The autonomous vehicles can cooperate to empower their cybersecurity. In future autonomous vehicles, Vehicle To Internet of Things, V2IOT, will be introduced within the clouds to reduce communication channels. Hence, this will further enhance the security by making targeting autonomous vehicle harder for attackers [37]. This collaborative defense that occurs in collaboration with cloud services will be part of cloud computing.

Passive defenses are carried out to detect, respond to, and recover from a security attack once it occurs. It can be summarized by finding ways to prevent malwares and code injection and modification techniques. Responding to these attacks to counteracting their impact is exercised using electronic or cyber capability, such as GPS anti-jamming device [38] or isolation. Isolation refers to detaching the autonomous vehicle from Inter-vehicle communication network to avoid harms to others.

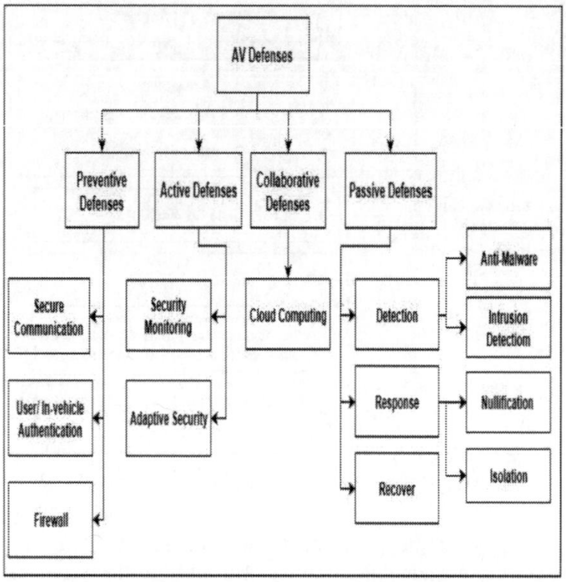

Figure 6. Defenses Types for Various Security Attacks

A number of European projects that are meant to enhance security techniques for the autonomous vehicle are ongoing. These include Secure Vehicle Communications (SEVECOM), Secure Hardware Extension(SHE), and E-safety Vehicle Intrusion Protected Applications (EVITA). SEVECOM works on enhancing the security of communications between autonomous vehicles. SHE aims at enhancing cryptographic processes to increase security [39].

Vehicular Ad-Hoc Networks (VANETs) is technique used for the Inter-vehicle communications. This can be protected against security threats regarding authentication and confidentiality due to the use of the digital signatures and private keys [40]. However, they still augment the possibility of vehicular botnets [31]. For this reason, researchers worked on improving the security for VANETs by verifying and certifying correctness of cryptographic authentication [41], analyzing messages sent by VANETs (in low level autonomous vehicles) [42]-[45], and analyzing trustworthiness of the sender [46].

VI. Conclusion

Vehicles were first introduced to provide transportation only for people. Through time and with the development of technology, vehicles were gaining more interest and are further improved. The goal of this improvement was to increase the security of people and this led to increased intelligence in data analysis within the vehicle itself. For this reason, autonomous vehicles were born. Nowadays, auto industry is capable of delivering level 3 autonomous vehicles. Some automakers are expecting full autonomous vehicle to be presented after couple years. Due to the increase usage of the communication networks within the autonomous vehicle, the vehicle is becoming more vulnerable to various security attack types. These attacks will have severe negative impact on the networks within the vehicle and could lead to disastrous incidents if attackers gain control over the autonomous vehicle. To overcome these attacks different defenses are combined with cybersecurity techniques. The purpose of these defenses is to ensure integrity, authenticity, and confidentiality of data transmitted within the autonomous vehicles. In particular, to prevent these attacks or at least minimize their impact, strong encryption and authentication need to be implemented. Intrusion Detection Systems (IDSs), and honeypots or honeynets should be considered. In parallel with these approaches, more smart sensors have to be introduced to replace the classical sensors. This will allow for cryptographic capabilities within these sensors as computing capabilities will be included.

References

[1] E. Eckermann, “World History of the Automobile,” SAE Press, pp. 14-14, 2001.

[2] H. Michelet, “L'inventeur Isaac de Rivaz,” pp. 1752 - 1828 Editions Saint-Augustin, (in French), https://books.google.com/books?id=Wf-nrnUaZxAC&pg=PA26&dq=François+Isaac+de+Rivaz#v=onepage&q=François%20Isaac%20de%20Rivaz&f=false [Retrieved: May, 2018].

[3] "DRP patent No. 37435," Archived from the original (PDF) on 4 February 2012, https://web.archive.org/web/20120204045616/http:/home.arcor.de/carsten.popp/DE_00037435_A.pdf, [Retrieved: May 2019].

[4] I. S. Jacobs and C. P. Bean, “Fine Particles, Thin Films and Exchange Anisotropy,” in Magnetism, vol. III, G. T. Rado and H. Suhl, Eds. New York: Academic, pp. 271–350, 1963.

[5] “Where to? A History of Autonomous Vehicle,” 2014, https://www.computerhistory.org/atchm/where-to-a-history-of-autonomous-vehicles/, [Retrieved, May 2019].

[6] M. Kane, "US Plug-In Electric Car Sales Charted," January 2019". http://InsideEVs.com, [Retrieved: May, 2019].

[7] J. Zhao and Q. C. B. Liang, "The Key Technology Toward the Self Driving Car," International Journal of Intellegent Unmanned Systems, vol. 6, pp. 2-20, 2018.

[8] J. Wang, J. Liu, and N. Kato, "Networking and Communications in Autonomous Driving, A Survey," in IEEE Communication Surveys and Tuorials, 2018.

[9] V. L. L. Thing and J. Wu, "Autonomous Vehicle Security: A Taxonomy of Attacks and Defences," in Proc. the IEEE International Conference on Internet of Things (iThings), IEEE Green Computing and Communications (GreenCom), IEEE Cyber, Physical and Social Computing (CPSCom), IEEE SmartData (SmartData), Chengdu, China pp. 164-170, 2016.

[10] E. B. Hamida, H. Noura, and W. Znaidi, “ Security of Cooperative Intelligent Transport Systems: Standards, Threats, Analysis and Cryptographic Countermeasures,” Electronics, vol. 4, pp. 380-423, 2015.

[11] National Highway Traffic Safety Administration, “Preliminary Statement of Policy Concerning Autonomous Vehicle,” https://www.nhtsa.gov/.../rulemaking/pdf/Automated_Vehicles_Policy.pdf , [Retrieved: May, 2019.

[12] K. Hyatt and C. Paukert “Self Driving Cars: A Level-By-Level Explainer of Autonomous Vehicle,” https://www.cnet.com/roadshow/news/self-driving-car-guide-autonomous-explanation/, [Retrieved: May, 2019].

[13] J. Short and D. Murray,” Identifying Autonomous Vehicle Technology Impacts on the Trucking Industry,” November 2016. http://atri-online.org/wp-content/uploads/2016/11/ATRI-Autonomous-Vehicle-Impacts-11-2016.pdf, [Retrieved: May, 2019].

[14] SAE International, “Automated Driving, Levels of Driving Automation are Defined in New SAE International Standard J3016,” 2014, http://www.sae.org/misc/pdfs/automated_driving.pdf, [Retrieved: May, 2019].

[15] I. Harner, “The 5 Autonomous Driving Levels Explained,” https://www.iotforall.com/5-autonomous-driving-levels-explained/, October 2017, [Retrieved: May, 2019].

[16] M. Burgess, “ When Does A Car Become Truly Autonomous? Levels of Self-Driving Technology Explained,” 2017, https://www.wired.co.uk/article/autonomous-car-levels-sae-ranking, [Retrieved: May, 2019].

[17] M. Burgess, " We Went Off-Road In Jaguar Land Rover's Autonomous Car," 2016, https://www.wired.co.uk/article/self-driving-autonomous-land-rover-jaguar-technology, [Retrieved: May, 2019].

[18] S. Lin, Y. Zhang, C. Hsu, M. Skach, E. Haque, L. Tang, J. Mars, " The Architectural Implications of Autonomous Driving: Constraints and Acceleration," in Proc. the Twenty-Third International Conference on Architectural Support for Programming Languages and Operating Systems (ASPLOS'18), Williamsburg, VA, USA, 2018, pp. 751-766.

[19] J. Stewart, "Tesla's Cars have driven 140M Miles on Autopilot. Here's How," https://www.wired.com/2016/08/how-tesla-autopilot-works/, [Retreived: May, 2019].

[20] Paul Godsmark, "The Definitive Guide to the Levels Of Automation for Driveless Cars," 2017. https://driverless.wonderhowto.com/news/definitive-guide-levels-automation-for-driverless-cars-0176009/, [Retrieved, May, 2017].

[21] M. Burgess, " Google has spun its self-driving car team out into a new company", December 2016, https://www.wired.co.uk/article/waymo-google-car-driving, [Retrieved: May, 2019].

[22] D. Fagella, "Self-driving Car Timeline for 11 Top Automakers," 2017, https://venturebeat.com/2017/06/04/self-driving-car-timeline-for-11-top-automakers/, [Retrieved: May 2019].

[23] Mobileye, "Mobileye C2-270 Essentials," 2017, http://prevenireaccidente.ro/brosuri/Mobileye%20C2-270%20Essentials%20Book%20-%20to%20print.pdf, [Retrieved: May, 2019].

[24] F. Lambert, "Tesla has a New Autopilot '2.5' Hardware Suite with More Computing Power for Autonomous Driving," https://electrek.co/2017/08/09/tesla-autopilot-2-5-hardware-computer-autonomous-driving/, [Retrieved: May, 2019].

[25] V. Nguyen,"2019 Audi A8 Level 3 Autonomy First-Drive: Chasing the Perfect 'Jam'," https://www.slashgear.com/2019-audi-a8-level-3-autonomy-first-drive-chasing-the-perfect-jam-11499082/, [Retrieved: May 2019].

[26] M. Mody, J. Jones, K. Chitnis, R. Sagar, G. Shurtz, Y. Dutt, M. Koul, M. G. Biju, and A. Dubey, "Understanding Vehicle E/E Architecture Topolgies for Automated Driving: System Partitioning and Tradeoff Parameters," in Proc. the Autonomous Vehicles and Machine Symposium, 2018, pp. 358(1)-358(5).

[27] J. R. Van Brummelen, M. O'Brien, D. Gruyer, and H. Najjaran, "Autonomoous Vehicle Perception System: The Technology of Today and Tomorrow", Trasportation Research Part C, vol. 89, pp. 384-406, 2018.

[28] R. Blake and M. Shiffrar, "Perception of Human Motion," Annual Review of Psychology, vol. 58, pp. 47-73, 2007.

[29] A. M. Wyglinski, X. Huang, T. Padir, L. Lai, T. R. Elsenbarth, and K. Venkatasbramanian, "Security of Autonomous Systems Employing Embedded Computing and Sensors," IEEE Micro, vol. 33, no. 1, pp. 80-86, 2013.

[30] M. T. Garip, M. E. Gursey, P. Reiher, and M. Gerla, "Congestion Attacks to Autoonomous Cars Using Vehicular Botnets," in Proc. the NDSS Workshop on Security of Emerging Network Technology (SENT'15), San Diego, CA, USA, 2015.

[31] M. T. Garip, P. Reiher, and M. Gerla, "Ghost: Concelaing Vehicular Botnet Communication in the VANET Control Channel," in Proc. the International Wireless Communications and Mobile Computing Conference (IWCMC), Paphos, Cyprus, 2016, pp. 1-6.

[32] S. Parkinson, P. Ward, K. Wilson, and J. Miller, "Cyber Threats Facing Autonomous and Connected Vehicles: Future Challenges," IEEE Transactions on Intelligent Transportation systems, vol. 18, no. 11, pp. 2898-2915, 2017.

[33] H. Hasbullah, I. A. Soomro, and J. B. A. Manan, "Denial of Service (DoS) Attack and its Possible Solution in VANET," International Journal of Electronics and Communication Engineering, vol. 4, no. 5, pp. 813-817, 2010.

[34] A. M. Jones, "Secure Isolation for Autonomous Vehicle Architecture," Future System Design, 2017, https://www.technologyscotland.scot/wp-content/uploads/2018/11/Secure-Isolation-for-Autonomous-Vehicle-Architectures.pdf, [Retrieved: May, 2019].

[35] L. Zussa, A. Dehbaoui, K. Tobich, J.M. Dutertre, P. Maurine, L. Guillaume-Sage, J. Clediere, and A. Tria, "Efficiency of a Glitch Detector against Electromagnetic Fault Injection," https://www.date-conference.com/files/proceedings/2014/pdffiles/08.3_1.pdf, [Retrieved: May, 2019].

[36] P. Luo, C. Luo, and Y. Fei, "System Clock and Power Supply Cross-Checking for Gitch Detection," https://eprint.iacr.org/2016/968.pdf, [Retrieved: May, 2019].

[37] A. Chattopadhyay and K. Y. Lam, "Autonomous Vehicle: Security by Design," ArXiv, 2018.

[38] Y. Cui and S. S. Ge, " Autonomous Vehicle Positioning with GPS in Urban Canyon Environments," IEEE Transaction on Robotics and Automation, vol. 19, no. 1, pp. 15-25, 2003.

[39] L. W. Li, L. Apvrille, and A. Bracquemond, "Design and Verification of Secure Autonomous Vehicles," in Proc. the 12th ITS European Congress, Strasbourg, France, 2017.

[40] A. Matar, M. Ashraf, and S. Nouh, "VANETS and Autonomous Driving," 2014, https://www.academia.edu/10109589/VANETS_and_Autonomous_Driving, [Retrieved: May 2019].

[41] J. Hubaux, S. Capkun, and J. Luo, "The Security and Privacy of Smart Vehicles", IEEE Security and Privacy, vol. 2, no. 3, pp. 49-55, 2004.

[42] M. El Zarki, S. Mehrotra, G. Tsudik, and N. Venkatasubramanian, "Security Issues in a Future Vehicular Network," European Wireless, vol. 2, 2002.

[43] P. Golle, D. Greene, and J. Staddon, "Detecting and Correcting Malicious Data in VANETs," in Proc. the 1st ACM International Workshop on Vehicular Ad Hoc Networks, pp. 29-37, New York, NY, USA, 2004, pp. 29-37.

[44] T. H. J. Kim, A. Studer, R. Dubey, X. Zhang, A. Perrig, F. Bai, B. Bellar, and A. Lyer, "Vanet Alert Endorsement Using Multi-Source Filters," in Proc. the 7th ACM International Workshop on Vehicular Internetworking (VANET'10), Chicago, Illinois, USA, 2010, pp. 51-60.

[45] T. Leinmuller, E. Schoch, and F. Kargl, "Position Verification Approaches for Vehicular Ad Hoc Networks", IEEE Wireless Communications, vol. 13, no. 5, pp. 16-21, 2006.

[46] F. Dotzer, L. Fischer, and P. Magiera, "Vars: A Vehicle Ad-Hoc Network Reputation System", in Proc. the Sixth IEEE International Symposium on a World of Wireless Mobile and Multimedia Networks, Taormina-Giardini Naxos, Italy, 2005, pp. 454-456.

Asynchronous Vehicle Control System Basing on Analytical Continuous-Time Functions

Damian Petrecki
Department of Computer Science and Management
Wroclaw University of Science and Technology
Wrocław, Poland
e-mail: damian.petrecki@pwr.edu.pl

Abstract — **The paper describes a proposition of a driver support system composed of multiple independent processes producing discrete outputs and consuming continuous inputs, with a shared interpolate process. The research rejects multiple controllers handling different areas with the same actuators in favor of single but both multi-criteria and asynchronous decision making system. This way, a decision making problem has been limited and a big data processing has been eliminated, keeping high vehicle performance and low physical system complexity. The solution presented in this paper offers very promising safety and comfort during the simulation-based experiments.**

Keywords – integrated driver support system; multi domain controller; continous-time controll; asynchronous algorithm.

I. Introduction

The paper describes a new way to decompose the driver support problem, not into stability control, anti-slip issue, extreme situation handling, etc., but into data acquisition, trajectory calculation and control execution, which provides comparable results: safe and comfortable ride. The solution is literally a heuristic algorithm performing the vehicle control task, basing on driver's reference input, exclusively producing actuator signals for all actuators in the systems. This way, a vehicle equipped with the proposed solution can uses a centralized computer system which eliminates multi-system interferences. Moreover, the vehicle can be easily maintain, including software based tuning, updating and introducing new features without adding new physical sensors and controllers or modifying existing ones. What is more, there is no ability to bypass the system by a driver, so it cannot be called a typical decision support system.

The paper structure is as follows. In section 2, the algorithm is compared to a classic approach. In section 3, the algorithm itself is described. In section 4, the simulation: in part A, the environment used during the research, in part B, both test end reference vehicles, in part C, performance evaluation, and in part D, dynamic experiments. In section 5, the results are evaluated. In section 6, a future development direction is shown.

II. State of the Art

It is hard to say what the most modern, scientific approach is for supporting drivers. There are well-known common safety systems, like Anti-Lock Braking System (ABS) [1] or Electronic Stability Control (ESC) [2], but the mainstream is developed under non-public licenses or even as companies' secrets. On the other hand, the most popular, related conference topics are vision and perception [3][4], traffic models [5][6], accident preventions [7][8], and, of course, autonomous driving [9]-[11].

This research presents a different perspective – it is an integrated driver support system, and the main goal is not developing a better perception system, a more precise model, or a smarter autonomous driver-replacement, but presenting a new way to compose different, existing solutions to achieve high performance (in a way of vehicle safety and comfort) while lowering the computing power at the same time.

Let us consider a case study, a very common situation, well know from everyday driving – a driver wants to launch rapidly with front wheels turned, similar to when entering the flow of traffic. A modern car, equipped with typical safety systems would involve a lot of these to influence the same parameter – wheels' speed. Engine Management System (EMS) [12] uses the engine to raise it, Acceleration Slip Regulation (ASR) [13] reduces it, active differential differentiates it, ECS applies brakes to avoid slipping and ABS limits this brake action. The proposed solution is a very different one. It would calculate the proper speed for all wheels, taking into consideration all variables handled by mentioned classic systems and apply it, in this very case even without using brakes, but only an engine and a transmission system.

What is more, classic in-vehicle systems often use very similar sensors, e.g., one camera for tracking traffic and the second one for analyzing traffic lanes [14]. The proposed solution aggregates data regardless of source, type and frequency and uses an assembled model, so it does not need the duplication and it is more hardware-independent, especially in case of frequency.

Current research is not related to an autonomous driving at all. The proposed system, thanks to its model of a

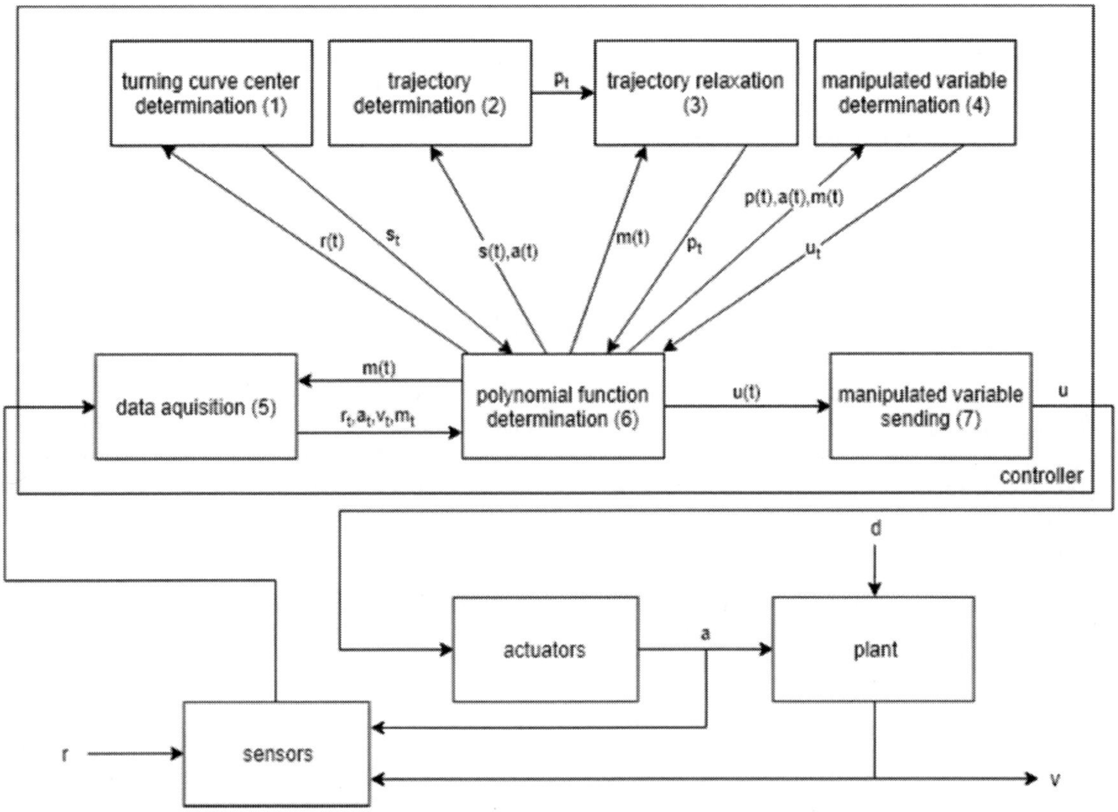

Figure 1. Control system schema

vehicle's environment could be a base of such solution, but for now this part is out of scope of the research.

III. ALGORITHM GENERAL DESCRIPTION

The algorithm is shown in Figure 1. When modeling using black-box method, ignoring the controller's structure, the presented solution seems to be very similar to typical control systems – it reads reference input *r* from the user (mainly steering wheel and pedals positions), the vehicle behavior in an environment (using cameras and radars) and vehicle-related data (using accelerometers, thermometers, etc.) *v* and also actuators state readers *a* and produces a control vector *u* consisting of all actuators manipulated variables: engine, linkage system, suspension, driveline, brakes. The only difference is the lack of time connection between inputs and the output.

The algorithm consists of several processes. Each of them can be scheduled (triggered by time) or started by data incoming from a sensor. The outputs of all processes are discrete values (in one or more dimensions) shown as sequence elements with bottom index *t*, e.g., s_t. Most input data (both starting processes and read during them) come from continuous-time functions stored in analytical, polynomial forms, shown as functions with *t*-argument, e.g., *s(t)*. It means a single, distinguished process (6) is introduced to build continuous-time functions from discrete sequences, which allows data interpolation and extrapolation. The process uses polynomial curve fitter method [15], accepts discrete values and timestamps, and produces a vector of polynomial coefficients. This way, a very specific storage is introduced that stores discrete variables and provides analytical functions as its output.

A single dimensional data acquisition process is proposed (5). It reads and stores input values from input devices r_t and in-vehicle sensors, it reads actuators' states a_t, like suspension status, accelerations, engine status, etc., and

TABLE I. SYMBOLS USED IN FIGURE 1

symbol	description
r	reference input form an user
a	actuators state
d	external distortions
v	behavior measured via sensors (cameras, radars, accelerometers, etc.)
m	environment map – list of measured objects with its position and classification and parametrization results
u	control vector
s	center of turn
p	vehicle position p_t/trajectory *p(t)*

simple (non-matrix) measurements from v_t. Each variable is handled by a separated thread.

The process of a second type calculates the desired trajectory using environment knowledge, vehicle-geometry model, and input data. It is split into several sub-processes, without any time-synchronization:

- The first sub-process referred to (1) in Figure 1 calculates the center of the turning curve (if any) in the vehicle-centered coordinate system, using speed, steering wheel position, and vehicle geometry.
- Sub-process (2) calculates the desired vehicle positions p_t in the future, which means the desired vehicle trajectory.
- Sub-process (3) uses genetic relaxation algorithm [16] and environment knowledge v_t to improve the trajectory to avoid accidents, lowering external objects hit possibility. Please note this process can change the trajectory in any way, e.g., by increasing speed or changing the turn, and its behavior is unpredictable. This is the only process that reads its input directly from the other process, not the storage.

The next process (4) uses the trajectory *p(t)* to calculate control values for all executors u_t, e.g., calculates each wheel speed and turn and then engine power, braking force, linkage system, and differential parameters. Calculated manipulated variables are being sent to the vehicle by own sender processes (7) (one process per variable), which read data from storage, not from the processes that actually generated them.

The process of the last type (5) handles the environment data v_t and is the most complex one. This is the complex part of the data acquisition process. This is the only case when the matrix data (distances from radars or bitmaps from cameras) have to be handled. The process is triggered for each input from each signal separately. The result is a 3D model of the vehicle's environment consisting of a set of classified objects, in the form of objects' shapes (3D line segments), class and positions, so data size is significantly decreased. Due to long processing time, the output of this process is stored with the input data appearance timestamps.

The environment analysis process is the biggest challenge related to the research. It needs a separate algorithm that accepts various formats of input data arriving at unpredictable time (cameras, radars) with, optionally, the already known 3D environment model *v(t)* to update the model as its output. The Long Short-Term Memory (LSTM) neural networks [17] are considered as the most promising way to solve this problem so far.

The most important idea in this algorithm is the absolute lack of time synchronization between input and

Figure 2. Simulator view (90-degree test)

output. Each process is run separately and uses functions extrapolated from other processes, no matter the age of the source values of the last polynomial calculation. It must be noted that all functions calculated during the process can be used by external processes, not related to vehicle control, like headlight control, climate control, comfort features, etc..

IV. CURRENT RESULTS

The presented solution has been tested in different scenarios and the current results are presented.

A. Simulation environment

All experiments are conducted in Unity3D [18] environment with Vehicle Physics Pro (VPP) [19]. Unity3D is responsible for communication with an operating system driver of Logitech G29 steering wheel [20], rendering visual interpretation of simulated rides and basic, Newton physics. VPP is responsible for vehicle simulation, including dependencies between in-vehicle physical subsystems, tires and suspension behavior, and standard active safety systems. VPP is in general a pre-compiled library, so a lot of its mechanisms are unknown. Its realism is not verified, just considered to be sufficient to compare two vehicles in the same conditions. The base assumption of the research is that the experiments results from a comparable, simplified environment should be applicable to the comparable, real one.

The experiments' results are read from telemetry panel provided by VPP (Figure 2) and from controller application. All data are stored during the experiments in text files and analyzed off-line.

B. Test and reference vehicles

The reference vehicle is built using VPP components only. It has an active suspension, 4-wheel steering, automatic gearbox, 4-wheel drive with active differential, and following active safety systems: ABS, Traction Control System (), ESC, ASR. Most of its implementation is hidden and unknown, but can be calibrated using built-in configuration panels, visible on the right part of Figure 2.

The test vehicle has the same physical parameters (weight 1200kg and equal weight distribution per wheel, wheels localizations, engine power/torque curves, etc.) and abilities (4-wheel drive, 4-wheel steering, controllable transmission, differential and suspension). The difference is that, in the test vehicle, the input from a driver is not sent to the vehicle itself, but transferred to an external application implementing the proposed algorithm. All in-vehicle and environment-related sensors data are handled in the same way. The application sends back control variable for each actuator separately, in separated threads. In this vehicle, there is no other driver support system implemented.

C. Performance results

All experiments are conducted using i7-7700k 4.2Ghz processor, 16GB RAM, SSD hard drive and Windows 10 64bit operating system. Both simulation (Unity3D) and control (external application) are performed on the same machine, because its performance is sufficient for current test scenarios. RAM usage never exceeds 10GB and CPU load is always below 20%, when simulation framerate 40fps is preserved.

The situation changes when all external sensors (21 radars and 8 cameras) are running. Then the simulation occupies about 4GB more RAM (which is still irrelevant), but exhausts all CPU abilities, reducing simulation framerate to 10-15fps (depending on a scenario).

D. Dynamic experiments

Three kinds of experiments are proposed. All test rides have been conducted 4 times, each with the same driver,

TABLE II. MOOSE TEST RESULTS

enter speed	reference solution			proposed solution		
	A	*B*	*C*	*A*	*B*	*C*
80	1.6	-4	170	0.1	2	91
80	1.8	-7	172	0.1	-1	84
80	1.7	-8	168	0.1	0	83
80	1.6	-5	192	0.2	1	86
100	1.6	-6	99	0.1	2	100
100	1.6	-6	97	0.2	11	87
100	1.7	-10	78	0.1	-2	81
100	1.7	-6	89	0.1	0	82
120	1.8	-9	145	0.2	1	92
120	1.5	-9	150	0.2	4	91
120	1.7	-2	154	0.2	4	97
120	2	-4	140	0.2	-1	95
140	failed			0.1	2	110
140	1.6	-40	180	0.4	2	101
140	2.1	-36	165	0.2	4	105
140	1.9	-38	79	0.3	1	87
160	1.9	-48	145	0.2	-2	89
160	1.8	-60	138	0.3	1	90
160	2.2	-40	139	0.3	2	98
160	failed			0.3	2	115
180	failed			0.4	2	97
180	2.1	-68	165	0.3	2	95
180	failed			0.3	1	96
180	failed			0.2	1	97
200	2	-20	66	0.2	2	94
200	Failed			0.3	1	119
200	1.9	-30	81	0.4	-6	126
200	failed			0.3	-1	83

same conditions. In Test 1 and Test 2, vehicle roll angle A (degree), speed change B (km/h), and maximum steering wheel angle C (degree) during the test are evaluated. Lower roll angle means better comfort. Lower loss of speed means higher safety (shorter maneuver time) and also better efficiency (energy loss). Lower steering wheel rotation angle is considered as sportier and also safer behavior, allowing driver to turn faster with holding the steering wheel with both hands all the time. In the last test, the most important evaluation parameter is minimum D and maximum E tires slip (m/s). Lower slip is equal to better handling and safer ride. The test is passed when the car fits the 4.3m lanes during the entire test.

1) Moose test

The test scenario is to rapidly change a lane and go back to the original one on a straight road with velocity in range 80-200km/h (changing by 20km/h).

Experiments results are shown in Table II.

2) 90-degree turn

The test scenario is to turn right on a 90-degree intersection with velocity in range 10-60km/h (changing by 10km/h).

Experiments results are shown in Table III.

3) Long curve

In this test, the vehicle rides around a circle with a constant radius of 20m and speed in range 10-70km/h (changing by 10km/h). In this test, the capability of controlling all wheel speed separately is shown.

Experiments results are shown in Table IV.

V. CONCLUSIONS

When analyzing results, some considerations arise. The reference vehicle has failed in 29% of all 1st and 2nd tests' trials and the tested one failed in 4% of the same tests' trials. This is the main proof of improved safety in the presented solution. Secondly, the maximum roll of test vehicle is limited to less than 0.5 degree no matter of conditions, for all

TABLE IV. 90-DEGREE TURN TEST RESULTS

enter speed	reference solution			proposed solution		
	A	*B*	*C*	*A*	*B*	*C*
10	0.2	2	253	0.2	-1	76
10	0.1	1	268	0.2	1	99
10	0.4	0	342	0.1	2	108
10	0.4	-2	268	0.2	0	104
20	0.8	1	372	0.2	-1	76
20	0.7	2	371	0.2	2	108
20	0.9	0	365	0.3	1	104
20	1	-1	312	0.2	0	98
30	1.2	2	290	0.2	2	96
30	1.2	1	246	0.2	1	94
30	1.1	-1	256	0.2	2	88
30	1.3	3	267	0.1	3	98
40	1.5	-10	160	0.2	-3	198
40	1.6	-8	381	0.2	-4	197
40	1.5	-12	271	0.3	-4	178
40	failed			0.4	3	174
50	1.9	-28	450	0.3	3	149
50	failed			0.2	-24	324
50	failed			0.3	-23	354
50	failed			failed		
60	failed			0.4	-38	450
60	failed			0.4	-39	450
60	failed			failed		
60	failed			0.4	-42	450

TABLE III. LONG TURN TEST RESULTS

enter speed	reference solution		proposed solution	
	D	*E*	*D*	*E*
20	0.01	0.23	0.07	0.07
20	0.02	0.25	0.06	0.07
20	0.01	0.22	0.05	0.07
20	0.01	0.23	0.06	0.06
30	0.06	0.4	0.1	0.11
30	0.05	0.42	0.09	0.11
30	0.06	0.41	0.09	0.12
30	0.07	0.4	0.11	0.13
40	0.18	0.72	0.27	0.31
40	0.2	0.7	0.22	0.25
40	0.19	0.7	0.25	0.3
40	0.17	0.71	0.27	0.3
50	0.22	1.81	0.2	0.21
50	0.21	1.9	0.23	0.26
50	0.24	1.86	0.25	0.29
50	0.24	1.78	0.21	0.24
60	0.25	2.59	0.26	0.37
60	0.25	2.72	0.25	0.28
60	0.27	2.58	0.25	0.27
60	0.26	2.58	0.26	0,29
70	0.26	2.42	0.21	0.25
70	0.25	2.44	0.22	0.25
70	0.26	2.53	0.21	0.26
70	0.26	2.51	0.25	0.28

trials. The reason is the anti-roll bars work pro-actively, reacting to turn and speed, not to the roll itself. This behavior proves that the comfort of the ride improved.

The next thing to notice is that the maximum steering wheel rotation angle in test vehicle is significantly lower and fits into 90 degree for most cases. The cause is the steering wheel ratio being adjustable in a very wide range, due to the lack of physical connection (even simulated one). The function converting wheel angle and speed to the position of the center of a turn is adjusted to lower the minimum turn radius at high speed, when rapid turns are impossible anyway due to vehicle momentum.

The next observation is that the presented vehicle does not slow down during most of the tests, except the ones, when preserving speed is impossible, due to high vehicle inertia. The cause is that the driver does not press the brake, so all trajectories are calculated for the same speed. Stability is preserved with an active differential that transfers proper speed to all wheels to avoid a slip, with stiff connection between wheels and engine and without using brakes. This way, a maneuver can be finished faster, and the engine is never stalled by brakes, which also improves safety.

On the other hand, the reference vehicle uses brakes to preserve comparable stability which causes a significant loss of speed.

In the last test, the minimum slip, occurring for inner wheels, is comparable for both vehicles, but active differential implementation offered by VPP is not as effective as the tested one. Moreover, in different conditions, when outer wheels travel on grass instead of tarmac, the differences between the reference and test vehicle are even bigger (Figure 3). The reason is, again, that tested standard safety systems react by breaking wheels already slipping, and the test one controls the wheels behavior proactively, calculating its speed before any slip occurs using mainly engine and differential, not brakes. This result also proves the higher ride safety.

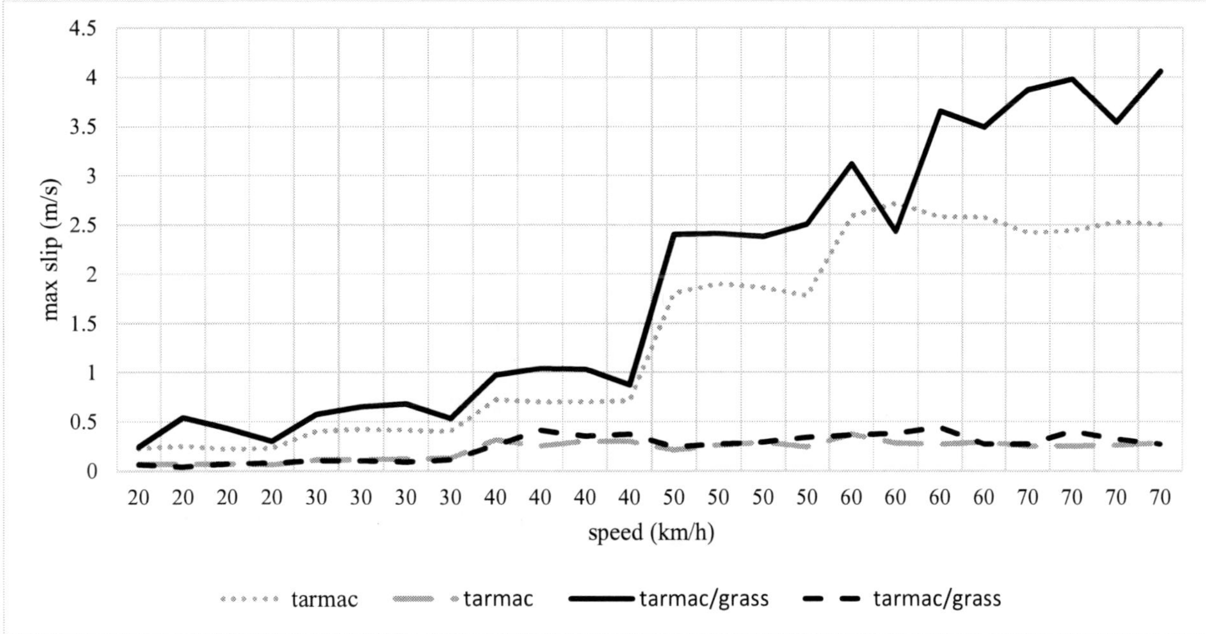

Figure 3. Slip test on different surfaces

VI. FUTURE WORK

Although the current results are promising, a lot of work is planned. For now, all processes are triggered by data or time. The event-base trigger (rapid condition change) is planned. Besides that, disruption analysis with fuzzy functions [21] usage will be introduced. When environment analysis process is established, the full evaluation will be conducted with more test cases and more drivers. And lastly, all processes implementations of the method use very simple algorithms so far, but they are designed to be replaceable, so the best combination is to be found.

Future experiments will be conducted using two computers with a direct network link. Physical experiments with real vehicles are not planned so far. Very sophisticated, highly equipped vehicle (with active differential, suspension, etc.) with an open access available to all in-vehicle actuators and sensors and also a large set of extra sensors are needed, which makes such experiments too expensive for the current stage of research. This kind of research is possible after full simulation evaluation.

REFERENCES

[1] J. Ernst, “Mercedes-Benz and the invention of the anti-lock braking system: ABS, ready for production in 1978”, in: Daimler Communications, July 2008.

[2] A. van Zanten, "Bosch ESP Systems: 5 Years of Experience" in: SAE Technical Paper 2000-01-1633, January 2000.

[3] M. Knorr, “Self-Calibration of Multi-Camera Systems for Vehicle Surround Sensing”, KIT Scientific Publishing, December 2018.

[4] B. Ranft, and C. Stiller, “The role of machine vision for intelligent vehicles”, in: IEEE Transactions on Intellifent Vehicles, Vol. 1, pp. 8-19, March 2016.

[5] M. Z. Liu, and M. Sun, “Application of Multidimensional Data Model in the Traffic Accident Data Warehouse”, in: Applied Mechanics and Materials, Volumes 548-549, pp. 1857-1861, April 2014.

[6] D. Badura, “Prediction of Urban Traffic Flow Based on Generative Neural Network Model”, in: Management Perspective for Transport Telematics. TST 2018. Communications in Computer and Information Science, volume 897, pp. 3-17, September 2018.

[7] Z. H. Ren, K. Zhang, L. X. Xue, and Y. L. Gao, “Mandatory Lane Change Model and Time Delay under Traffic Emergency Incidents”, in: Applied Mechanics and Materials, Vols. 644-650, pp. 2627-2631, September 2014.

[8] H. Abut, J. Hansen, G. Schmidt, K. Takeda, and H. Ko, “Vehicle Systems and Driver Modelling”, De Gruyter, September 2017.

[9] T. Nitsch, “Sensor Systems and Communication Technologies in Autonomous Driving”, GRIN Verlag, April 2017.

[10] S. Liu, L. Li, J. Tang, S. Wu, and J. Gaudiot, “Creating Autonomous Vehicle Systems”, Morgan & Claypool Publishers, October 2017.

[11] D. Vaishnavi, E. Sundari, T.V. Sangeetha, S. Shrinidhi, and P. Saravanan, “Design and Development of Computational Intelligence for Enhanced Adaptive Cruise Control Using Arduino”, in: Applied Mechanics and Materials, Vol. 852, pp. 782-787, September 2016.

[12] T. Denton, “Automobile Electrical and Electronic Systems”, Chapter 10, Routledge, July 2007.

[13] D. Hoffman, "The Corvette Acceleration Slip Regulation (ASR) Application with Preloaded Limited Slip Differential," SAE Technical Paper 920642, February 1992.

[14] E. D. Dickmanns, “Dynamic Vision for Perception and Control of Motion”, chapter 7, Springer, June 2007.

[15] P. G. Guest, “Numerical Methods of Curve Fitting”, Cambridge University Press, December 2012.

[16] C. Cheng, C. Liu, and C. Liu, “Unit Commitment by Lagrangian Relaxation and Genetic Algorithms”, in: IEEE Transactions on Power Systems, Vol. 15, pp. 707-714, May 2000.

[17] A. Yenter, “A Multi-kernel Convolutional Neural Network with LSTM for Sentimental Analysis”, ProQuest, 2017.

[18] https://unity3d.com/, last retrieved May 2019.

[19] https://vehiclephysics.com/, last retrieved May 2019.

[20] https://www.logitechg.com/pl-pl/products/driving/drivingforce-racing-wheel.html, last retrieved May 2019.

[21] J. J. Buckley, and E. Eslami, “Fuzzy Functions” in: An Introduction to Fuzzy Logic and Fuzzy Sets. Advances in Soft Computing, Physica, 2007.

Robustness Against Hazard Notifications Around a Vehicle Using Seat Actuators

Akimasa Suzuki
and Yoshitoshi Murata
Department of Information Science,
Iwate Prefectural University
020-0693, Sugo, Japan
suzuki_a@iwate-pu.ac.jp

Shoma Fujimura
Ad-Sol Nissin Corporation,
Kawasaki-shi, 210-0804,
Kanagawa, Japan
oaurn85ns@outlook.jp

Abstract—**This paper examines the robustness of our proposed haptic notification system against the different types and layers used for driving seat cushions. While many car manufacturers provide useful side and rear collision warning systems with sound alarms or visual monitors, the addition of similar notifications can confuse a driver because they already need to be aware of many visual targets such as mirrors, monitors, and environmental sounds. Therefore, we have investigated a haptic notification system that uses the driver's buttocks. The results show that drivers can correctly identify the directions of five vibrating motors, three intensity settings, and three obstacle types (*i.e.*, pedestrians, vehicles, and motorcycles). In this paper, we investigate whether drivers can discriminate the direction, intensity of vibrations, and vibration patterns of the system through their buttocks to identify the obstacle direction, degree of risk, and the type of obstacle, even if the vibrations are attenuated by the seat cushion. The results indicate the high potential of the haptic sensation system to notify the driver of obstacles, especially those located in the blind spot.**

Keywords-Vibro-Tactile Notification; Type of Obstacle; Buttocks; Acoustic Haptic Actuators; Seat Cushion.

I. INTRODUCTION

There has been considerable research in investigating accident prevention systems for vehicles, particularly in relation to developing driving support systems that will transition to autonomous driving systems. However, to realize autonomous driving systems, we must overcome problems related to cyber-security measures and traffic laws (*e.g.*, responsibility for accidents by autonomous cars [1]), which could take time. Additionally, as many people enjoy driving, the demand for manual driving as a hobby is unlikely to fade. Driving support systems will thus remain an important feature. Moreover, despite the high number of driving support systems used in Japan, many fatal vehicle accidents are caused by violations of safe driving practices, such as failing to keep eyes on the road, careless driving, and failing to make safety checks [2], thus highlighting the need to develop more techniques that support drivers.

To develop a support system that helps drivers to avoid vehicle accidents, the system needs to quickly and accurately sense information and notify the driver so that he/she can make a rapid judgement. Most car manufacturers now install highly accurate sensor systems at the front and rear of their vehicles at a low cost. Support systems located at the front of a vehicle use vision [3] or radar [4] sensors to prevent careless driving and overcome a driver's failure to make safety checks, while support systems located at the rear of a vehicle use sensors and notification systems to monitor a driver's rear view and blind spot [5]. These systems use sound or visual images to alert drivers to potential hazards.

Visual images can quickly notify a driver about many kinds of information using shapes and colors. As vision is the dominant human sense [6], many notifications rely on the driver's vision, including the front view, mirror, tachometer, speedometer, navigation system, and indicators. There is a concern, therefore, that excessive visual information could affect a driver's capacity to adhere to safe driving practices [2]. We thus consider that developing an additional visual notification may cause the driver to confuse it with conventional visual notifications.

Many conventional systems also provide information to drivers in the form of sounds (*e.g.*, alerts by horn; car audio, including radio; and alarms for reverse gear, pre-collision, and lane departure). Directions presented by a satellite navigation system are also expressed through the vehicle's stereo system. To avoid confusing the driver, we considered creating different sounds, pitches, and patterns for each type of obstacle; however, these would not be intuitive. Additionally, notifying a driver using speech would be too slow to get communicate the message in time. It is also difficult to apply a system using sound on a late-night bus travelling long distances because sleeping passengers may get up by the alert.

Therefore, we proposed a system that uses haptic sensations to quickly notify drivers of possible hazards or obstacles surrounding the vehicle [7]. Our proposed system has higher immediacy and directional resolution than notifications using sounds. As no driver notification system currently uses haptic sensations, we do not have to consider conflicts in this area. Our proposed notification system uses vibro-tactile haptic devices that remain in constant contact with the driver's buttocks. We evaluated the system's robustness against cushion type for determining the direction and intensity of vibrations and road conditions. A high intensity expresses the extent of the danger and the direction of the vibration indicates the location of the hazard. The system can also alert the driver to different types of obstacles, such as a pedestrian, car, or bike. The results indicated a high potential for notifying drivers of obstacles, especially those located in the blind spot.

To support safe car driving, our proposed haptic notification system installing vibration alerts into a driving seat. This paper examines its robustness against different types and layers used for driving seat cushions.

The remainder of the paper is structured as follows. Section II discusses relevant studies, Section III describes the proposed system, Section IV describes the modulated waveforms generated for precise notification, Section V presents the experiments to test the robustness of the system, and Section VI presents our conclusions.

II. Relevant Studies

Many practical driving support systems apply image sensors [3], radar [4], and ultrasonic sensors [8] to detect pedestrians and other vehicles with high accuracy. Around-view monitors are increasingly being used for automatic parking [9] and lane-detection systems are being applied using three-dimensional (3-D) laser imaging detection and ranging (LIDAR) [10]. Despite their weakness to other noise sound, ultrasonic sensors can now be installed in driving support systems for a low cost, while the cost of 3-D LIDAR is also dropping. These devices can be used to detect not only the presence of an obstacle, but also the type of obstacle (*e.g.*, pedestrian, vehicle, or motorcycle). However, in this research, we focus on creating a notification method to alert drivers to the potential hazards, rather than the development of a sensor system.

Previous research on evaluating seat comfort has demonstrated that buttocks are sensitive to tactile sensations [11]. Although not used in the driving seat, some studies have reported the effectiveness of vibro-tactile devices for notifying drivers of directions when using a wearable device such as a belt [12]. The directions of obstacles could be detected by using vibro-tactile devices on the seatback [13] because the back is more sensitive than the buttocks; however, as drivers need to lean against the backrest, the system might have a negative effect on the driver's posture. A vehicle notification device using vibro-tactile devices on the buttocks was therefore developed [14], although the system was unable to indicate the direction of a hazard to the driver.

In a gaming device, vibro-tactile devices are used to link a virtual object with reality [15]. Therefore, we consider applying a vibro-tactile device to notify the driver of essential information related to potential hazards based on the intensity and direction of vibrations. Tactile sensations can include rubbing, pain, pressure, and warmth. On the streets, tapping on the shoulder is a popular method for pedestrians to alert each other. To our knowledge, this study is the first to notify a driver of information such as the direction of a hazard in relation to the vehicle, the extent of the urgency based on the intensity of the vibration, and the type of hazard by the vibration pattern expressed using vibro-tactile devices located below the buttocks.

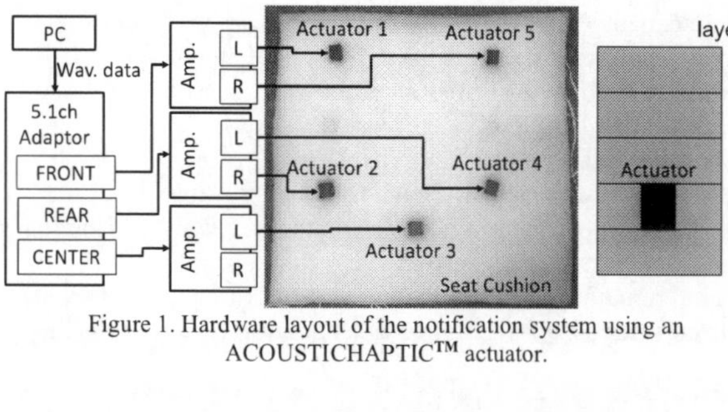

Figure 1. Hardware layout of the notification system using an ACOUSTICHAPTIC™ actuator.

III. Vibro-tactile Notification System Around a Vehicle Using Seat Actuators

We will indicate our proposed vibro-tactile notification system in this section.

A. System Architecture

For our proposed system, we utilized a vibrating motor with an ACOUSTICHAPTIC™ actuator developed by Foster Electric Company Limited. The acoustic haptic actuator is a kind of woofer that comes into direct contact with the driver's buttocks. Fig. 1 shows the hardware layout for this system. The edited waves were played on a PC, and the five actuators vibrated on the seat, as shown in Fig. 1. These actuators contact with the back of the driver's knees. We used the AP05 amplifier produced by Fostex.

In this experiment, we administered four vibrating patterns of the same intensity, representing different obstacle types, to fifteen participants. The participants were asked to identify the type of obstacle from the vibration pattern. We conducted five trials in a random order for each participant. The vibrations included the sound of footsteps from leather shoes [16], the sound of a V6 engine revving up [17], the sound of a bus driving uphill [18], an idling sound [19] as obstacle types of pedestrian, small and large four-wheeled vehicles, and a motorcycle, respectively. We hypothesized that drivers would intuitively recognize the type from the vibration pattern of the real sound.

As shown in Fig. 1, up to three layers of urethane cushions were placed over the actuator to evaluate the robustness. The thickness of each layer is 2 cm. We also define "layer 0" to mean that nothing is placed over the actuators. We utilize three types of urethane cushion, with specifications shown in Table 1. We defined 20 ss, 35 s, and BZ-10 constructed by Toyo Quality One as soft, highly resilient, and less resilient cushions, respectively, as shown in Table 1.

We generated vibration waveforms from these sound data, to decreasing up to 2 kHz and increasing between 55 Hz and 110 Hz, which are resonance frequencies of the ACOUSTICHAPTIC™ actuator. Fig. 2 shows the waveforms of these four vibrating patterns, *i.e.*, (a) a pedestrian, (b) a

TABLE I. CUSHION MATERIALS FOR EVALUATION.

	Soft	**Highly Resilient**	**Less Resilient**
Constructor	Toyo Quality One	Toyo Quality One	Toyo Quality One
Product name	20 ss	35 s	BZ-10
Density (kg/m^3)	20 ± 2	55 ± 2	35 ± 3
Hardness (N)	30 ± 15	45 ± 15	60 ± 15
Tensile intensity (kPa)	$50 \leq$	$60 \leq$	$30 \leq$
Elongation (%)	$200 \leq$	$100 \leq$	$80 \leq$
Tensile intensity (N/cm)	$3.0 \leq$	$2.0 \leq$	$2.0 \leq$
Compressive residual strain (%)	$10 \geq$	$12 \geq$	$15 \geq$

motorcycle, (c) a small 4-wheel vehicle, and (d) a large 4-wheel vehicle. The horizontal and vertical axes indicate the time and amplitude, respectively. A waveform of the pulse vibration with a walking frequency of 0.4-s intervals was used for the pedestrian. We applied the 55-Hz and 110-Hz resonance frequencies to large and small four-wheeled vehicles, respectively. The amplitude of the waveforms was normalized because we utilized the different amplitudes (*i.e.*, the intensity of the vibration) to express the urgency of the degree of risk or the distance to the obstacles.

B. Abilities of the Proposed Notification Systems

A study of our proposed conventional system proved it to be effective [7]. However, the vibrating motor used in our conventional system was unable to assign different vibration patterns to different obstacle types. Fig. 3 shows the correct answer rates for (a) the direction, (b) vibration intensity, and (c) both direction and intensity using the vibrating motor. When participants took longer than 5 s to respond, measured using a stopwatch, we considered it to be too slow and treated their answer as incorrect. In the correct answers shown in Fig. 3, the response times of all trials indicate that drivers could understand the information of the surrounding obstacles in less than 1 s. The correct answer rates for direction, intensity, and both direction and intensity in all route types were 84.4%, 72.6%, and 62.2%, respectively [7].

Fig. 3(a) shows that the drivers produced the highest number of correct answers when driving on the winding local road, followed by the arterials and the collector roads; however, as Fig. 3 (b) shows, the accuracy of the drivers' responses for the intensity were in the reverse order. This difference is likely because the vehicle's vibrations when travelling at low speeds could confuse the driver. The pressure between the vibrating motor and the buttocks could also change during the trials because the driver had to constantly control the accelerator and brake on the winding road. Nevertheless, the participants were able to determine the direction and intensity of over 50% of the vibrations when driving on the proposed seat.

For determining obstacle types, we applied the ACOUSTICHAPTIC™ actuator. Figs. 4 and 5 indicate the correct answer rates for the four types of obstacles. The graph also shows the rate for each trial by type. The vertical and horizontal axes in Figs. 4 and 5 show respectively the correct answer rate and vibration type of the waveforms shown in Fig. 2. Fig. 4 presents the correct answer rates based on the waveforms of the four obstacles types. As Fig. 4 shows, the correct answer rates improved during the trial, except for those for the four-wheeled vehicle (small). All participants were able to identify the pedestrian and motorcycle vibrations; however, they could only identify 50% of the other types of vibrations because the vibration patterns were too similar for them to sense the differences. However, after combining the large and small four-wheeled vehicles, the participants could detect the three patterns with high accuracy. Fig. 5 shows the results for three obstacle types, integrating the large and small four-wheeled vehicles. The correct answer rate reached over 90% at the fifth trial, as shown in Fig. 5.

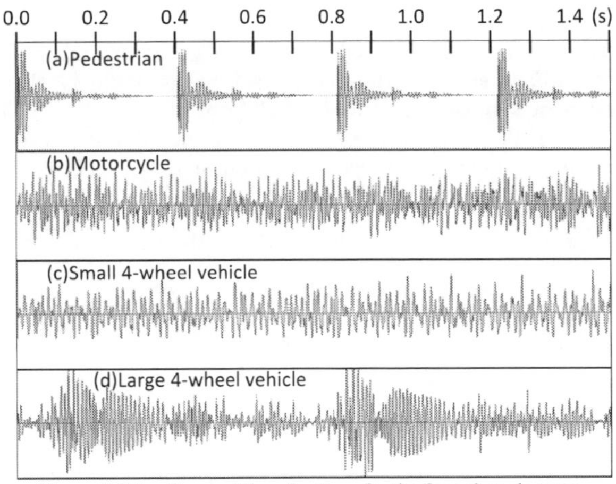

Figure 2. Waveforms of the vibrations for the four obstacle types.

IV. MODULATION FOR PRECISE NOTIFICATION

From the waveforms shown in Fig. 2, we generated modulated waves for precise notification. We determined the waves with a frequency an octave lower than the original wave as the modulated waves for more clearly feeling the differences of vibration. Figs. 6(a) and (b) show spectra of the original and modulated waves for large four-wheeled vehicles. The horizontal and vertical axes indicate the frequency and power spectrum, respectively. The spectra of the modulated waves consist of sine waves under 1 kHz.

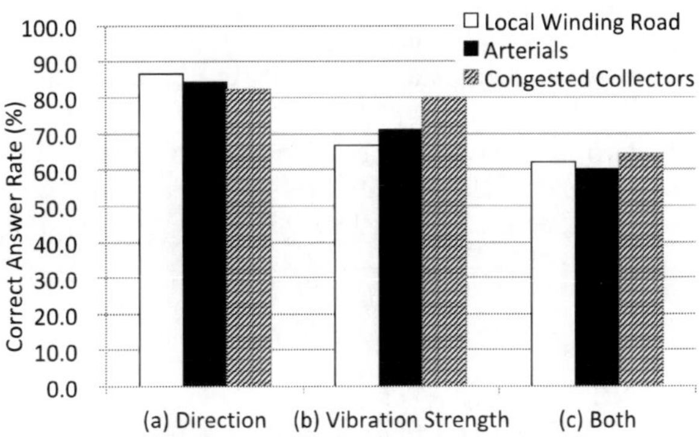

Figure 3. Correct answer rates on local winding roads, arterials, and congested collectors.

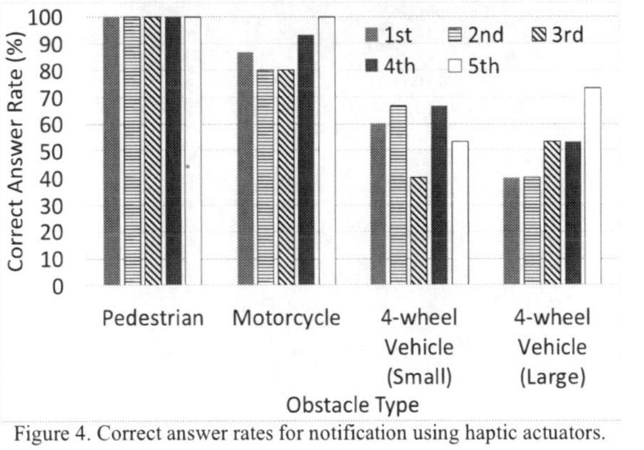

Figure 4. Correct answer rates for notification using haptic actuators.

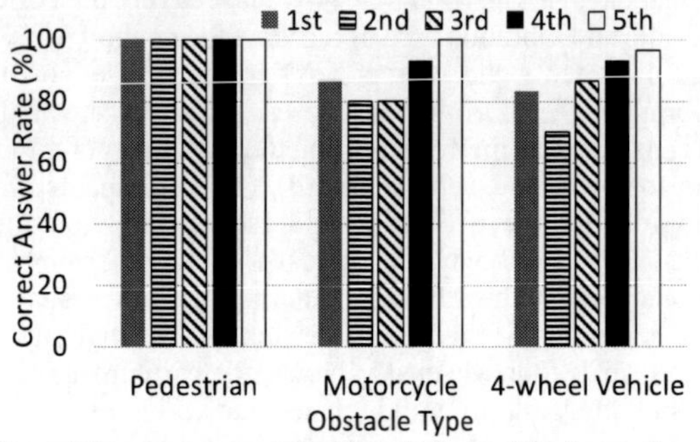

Figure 5 Correct answer rate for obstacle type notification on three obstacle types

For the intensity expression, we utilized three intensity waves, as shown in Table 2, which shows three vibrating volumes corresponding to three steps of intensity (*e.g.*, small, medium, and large). We applied 8 dB intervals between the three steps, and we prepared the signals with the three steps of volume on each modulated waveform for the three obstacle types: pedestrians, motorcycles, and four-wheel vehicles.

V. EXPERIMENTS FOR ROBUSTNESS AGAINST CUSHION TYPE

We evaluated the robustness against cushion type in a near-practical environment. As shown in Fig. 1, we mounted a vibrating car seat on a test vehicle for evaluation by five test drivers with considerable driving experience. The test drivers reported the vibration intensity, direction, and obstacle type when they sensed the vibration. Before the evaluation, the test drivers felt nine types of vibrations (*i.e.*, three intensities for the three obstacle types) at each actuator, shown as Actuator 1 to Actuator 5 in Fig. 1. The bold line in Fig. 7 indicates the experimental route. The actuator is vibrated at random times while the test drivers drive on a circuit track, shown in Fig. 7, at speeds of less than 20 km/s. Answers were only considered valid when received within 5 s of the vibration.

TABLE II. SOUND VOLUME CORRESPONDING TO THREE STEPS OF VIBRATION INTENSITY.

	Volume (dB)
Small	−16
Medium	−8
Large	0

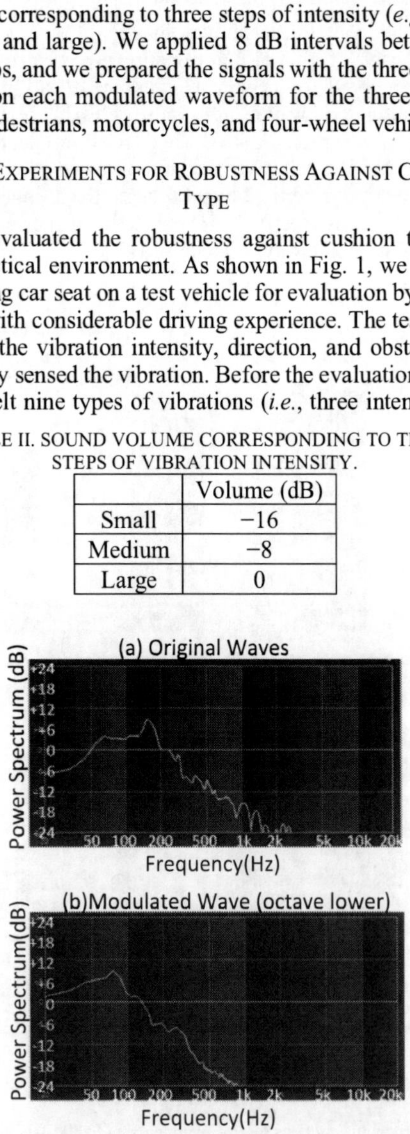

Figure 6. Sound spectra for 4-wheeled vehicles (a) before and (b) after modulation.

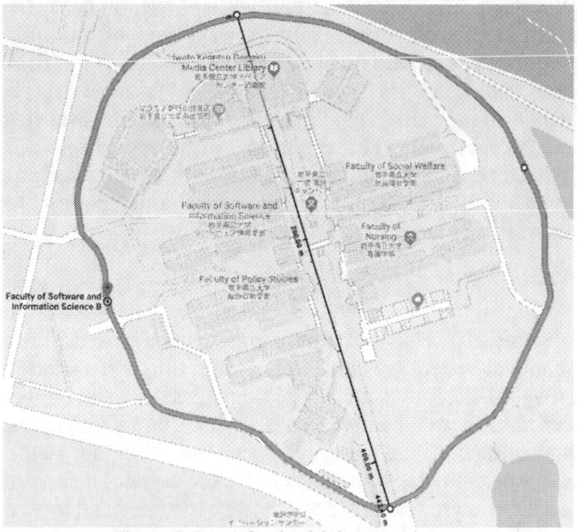

Figure 7. Route for the driving experiment.

A. Experimental Results

Fig. 8 shows the average ratings for comfort of each seat layer and material. The vertical and horizontal axes represent the average comfort rating and the cushion layer and material, respectively. The ratings were ranked according to the softness and resilience (high or low) of the seat cushion up to several layers. The test drivers tended to evaluate the seat based on whether they were conscious of the actuators.

The experimental results indicate the importance of retaining a high notification ability in a thick cushion even though a synthetic judgment is required for other evaluations, such as ease of driving.

Figs. 9 to 12 show the results for robustness against cushion type and present the correct answer rates as relative values based on a correct answer rate for layer 0. The vertical and horizontal axes present the correct answer rate and cushion layer and material, respectively. Figs. 10 to 12 also present the standard deviations for all answers.

Fig. 9 shows the differences between the correct answer rates for the intensity, direction, and obstacle types between

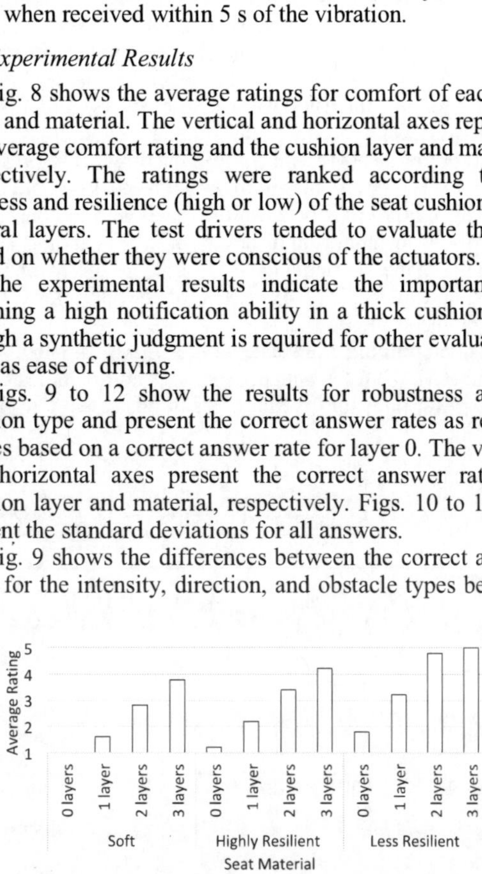

Figure 8. Comfort ratings for each seat layer and material.

layer 0 and the other layers. The results show that the correct answers for the highly resilient cushion decrease as the layers increase; however, the other cushion materials maintain robustness even with an increased cushion thickness.

B. Intensity Expression

Fig. 10 shows the average differences in the correct answer rates calculated from the answers relating to the intensity of the vibrations (*e.g.*, small, medium, or large, as shown in Table 2). The more layers the seat cushion has, the more the correct answer rates decrease, except for the less resilient cushion. A high standard deviation is obtained for the results of the less resilient cushion, although the trend of the correct answer rates for the layers is different. Therefore, it cannot be said that the number of correct answers will increase for more layers in the less resilient cushion.

C. Direction Expression

Fig. 11 shows the average differences in the correct answer rates, which are calculated from only the answers relating to direction (*e.g.*, left corresponding to Actuator 1 and right corresponding to Actuator 5, shown in Fig. 1). In this experiment, the test drivers gave the direction by stating "right", "right back", "back", "left back", or "left" when they noticed the vibration. The results shown in Fig. 11 confirm substantial differences between the different seat cushions.

D. Obstacle Type Expression

Fig. 12 shows the average differences in the correct answer rates calculated from the answers relating to the obstacle types (*e.g.*, pedestrians, motorcycles, or four-wheeled vehicles, shown in Fig. 5). The more layers there are, the more the correct answers decrease, except for the soft cushion. In the case of the soft type with 0 layers, it was difficult to judge when the actuators made direct contact with the buttocks based on the features of the waveform because the soft type of seat sank more easily than the other types. As a result, the correct answer rates on the soft cushion could be increased.

Based on the results shown in Figs. 9 to 12, the total correct answer rates are strongly influenced by the intensity of the vibrations; thus, the robustness is demonstrated without to the intensity steps of the vibrations.

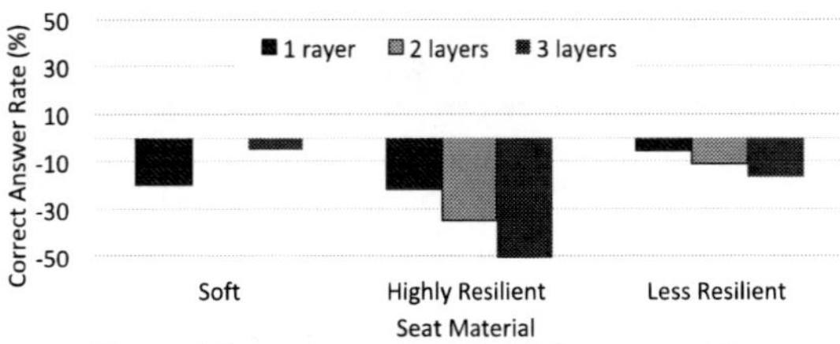

Figure 9. Correct answer rates for each seat material.

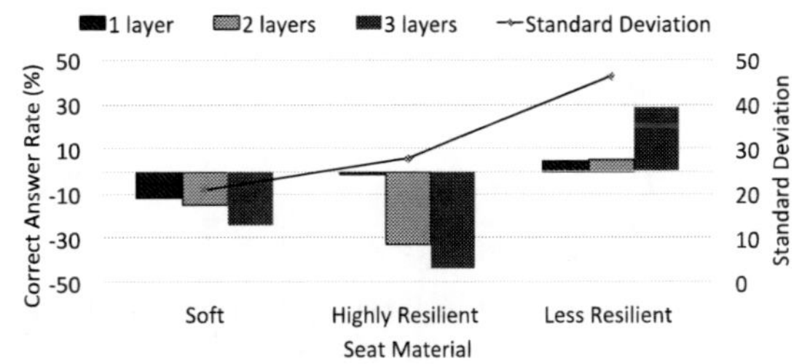

Figure 10. Correct answer rates for intensity for each seat material.

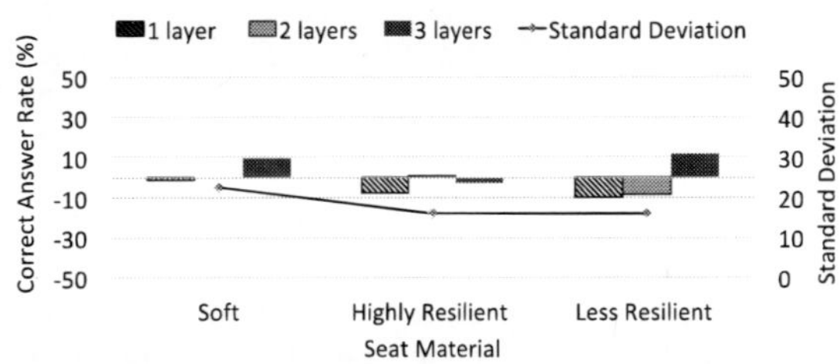

Figure 11. Correct answer rates for direction for each seat material.

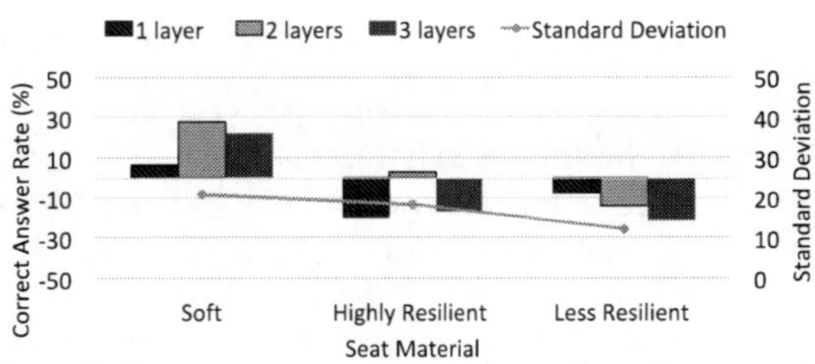

Figure 12. Correct answer rates for obstacle type for each seat material.

VI. CONCLUSION

We examined the robustness of a vibro-tactile device by collaborating with a car manufacturer to install acoustic haptic actuators into the seat cushion of an actual automotive vehicle.

We proposed a vibro-tactile notification system using vibrating motors to notify drivers of hazards around a vehicle, which were sensed using conventional sensors. The effectiveness of this method was evaluated from the viewpoint of resolution of intensity and direction and robustness against cushion type for determining the road conditions. The vibration pattern also enabled drivers to recognize the type of hazard, such as an approaching pedestrian or motorcycle.

We conducted several experiments involving driving on public roads in a car with seven vibrating motors installed under the driver's seat. By applying acoustic haptic actuators as a vibro-tactile device, test drivers could detect three types of vibrating patterns, indicating different types of obstacles: pedestrians, motorbikes, and four-wheeled vehicles. We also determined that drivers' recognition of the intensity, direction, and hazard type could be improved over time because they could learn from experiencing the vibration alerts.

The results indicated the high potential of using a haptic sensation device to notify drivers of obstacles in their blind spots by creating a vibration against the buttocks. The experimental results, shown in Fig. 9 to 12, illuminated that the intensity of the vibration, which should indicate the level of the hazard, could not be considered in the robustness test. By reconfiguring the intensity, as shown in Table 2, the robustness could be improved, which we will investigate in future work.

The proposed system is expected to reduce accidents by notifying drivers of other drivers and obstacles. In future works, we will conduct these experiments using more test drivers to compare elderly people with very young people or to observe its effectiveness with truck drivers and people with different levels of attention or tiredness. We will also evaluate the operational difficulties of the system in case of an emergency. We will also corroborate the visual and audio notifications of our system and examine the effectiveness for making quick driving decisions with intuitive notifications.

ACKNOWLEDGMENTS

This work was supported by MEXT KAKENHI Grant Number JP 16723884 and Foster Electric Company, Limited.

REFERENCES

[1] A. Hevelke and J. Nida-Rümelin, "Responsibility for crashes of autonomous vehicles: an ethical analysis," Science and Engineering Ethics., vol. 21, no. 3, pp. 619–630, June 2015.

[2] National Public Safety Commission and National Police Agency, "Traffic accidents situation (2017Oct.)," Dec. 2017. [Online]. Available from: http://www.e-stat.go.jp/SG1/estat/GL32020101.do?method=xlsDownload&fileId=000008051596&releaseCount=1 2019.06.24

[3] Subaru of America, Inc., Dec. 2017. [Online]. Available from: https://www.subaru.com/engineering/eyesight.html 2019.06.24

[4] Toyota Motor Sales, U.S.A., Inc., "SUZUKI GSX-S1000 normal idling sound motorcycle," Dec. 2017. [Online]. Available from: https://www.toyota.com/safety-sense/animation/pcs 2019.06.24

[5] Infineon Technologies AG, Dec. 2017. [Online]. Available from: https://www.infineon.com/cms/en/about-infineon/press/market-news/2009/INFATV200905-058.html?redirId=46231 2019.06.24

[6] I. Rock and J. Victor, "Vision and touch: an experimentally created conflict between the two senses," Science, vol. 143, no. 3606, pp. 594–596, Feb. 1964.

[7] A. Suzuki, Y. Murata, and M. Hayashi, "Notification of hazards around a vehicle using seat actuators," in 25th ITS World Congress, Copenhagen, Denmark, 17–21 Sep. 2018.

[8] L. Akonso, V. Milanés, C. Torre-Ferrero, J. Goboy, J. P. Oria, and T. de Pedro, "Ultrasonic sensors in urban traffic driving-aid systems," Sensors, vol. 11, pp. 661–673, Sep. 2011.

[9] J. K. Suhr and H. G. Jung, Fully-automatic Recognition of Various Parking Slot Markings in Around View Monitor (AVM) Image Sequences, 15th International IEEE Conference on Intelligent Transportation Systems (ITSC), Anchorage, USA, 1244-1299, Sep. 2012.

[10] L. Qingquan, C. Long, L. Ming, S. Shih-Lung, and N. Andreas, "A sensor-fusion drivable-region and lane-detection system for autonomous vehicle navigation in challenging road scenarios," IEEE Vehicular Technology Society, vol. 63, no. 2, pp. 540–555, Feb. 2014.

[11] M. Tada, S. Sekiguchi, T. Nisimatsu, and E. Toba, "Measurement and evaluation of sitting comfort of automotive seat," in Instrumentation and Measurement Technology Conference, 1998. IMTC/98. Conference Proceedings. IEEE, St. Paul, MN, USA, USA, 1998.

[12] A. Cassinelli, C. Reynolds, and M. Ishikawa, "Augmenting spatial awareness with the haptic radar," i10th IEEE International Symposium on n Wearable Computers, Montreux, Switzerland, 2006.

[13] R. W. Lindeman, Y. Yanagida, J. L. Sibert, and R. Lavine, "Effective vibrotactile cueing in a visual search task," IFIP TC13 International Conference on Human–Computer Interaction (INTERACT '03:), Zurich, Switzerland, 2003.

[14] C. C. Ltd., "Vehicle notification device, vehicle notification method and notification signal". Japan Patent WO2017090355 A1, 1 6 2017.

[15] Alps Electric Co., Ltd., Dec, 2017. [Online]. Available from: http://www.alps.com/prod/info/E/HTML/Haptic/ 2019.6.24

[16] M.-studio. official Channel, "Recommend sound effect, Footstep-Leather shoes 04 On-Jin" [Online]. Available from: https://www.youtube.com/watch?v=A5fGVbsHOfI 2019.06.25.

[17] GYANTARO, "Vellfire V6 3.5L 0-138km/h Full Acceleration Test (Alphard 30 series)," [Online]. Available from: https://www.youtube.com/watch?v=WI4voS7YF6Y 2019.06.25.

[18] TAKAPON CHANNEL, "Saihi tourist bus JAPAN" [Online]. Available from: https://www.youtube.com/watch?v=EsJnG-E22ss 2019.06.25.

[19] ahotakuro, "SUZUKI GSX-S1000 Normal Idling Sound Motorbike" [Online]. Available from: https://www.youtube.com/watch?v=c6pK2z9ANRk 2019.06.25.

[20] Google, [Online]. Available: https://www.google.com/maps/@39.7560755,141.1615498,13.42z 2019.06.25.

Vehicle Detection Assistance in Urban Intersection Using Data Exchange Between Road Infrastructure

Jean Marchal, Denis Gingras
Laboratory on Intelligent Vehicles
Engineering department, University of Sherbrooke
Sherbrooke, Québec, Canada
Email: jean.marchal@usherbrooke.ca
Email: denis.gingras@usherbrooke.ca

Hervé Pollart
Opal-RT ITS
Opal-RT Technologies
Montréal, Québec, Canada
Email: herve.pollart@opal-rt.com

Abstract— **This paper provides a new way of handling results related to safety algorithms used in urban area, in an intersection network. It denounces the waste of computing resources granted to run those algorithms while most mobile entities targeted may be detected several times during their journey through multiple interconnected intelligent intersections. A new approach is proposed, mixing detection algorithms and communications between intersection in order to reinforce consecutive detections of the same vehicles at different places. Experimentation in simulated environment leveraging a vehicular simulator called SCANeR Studio give some payoff on its performances: some improvements relative to the precision of vehicle detections but also a slight shift in detected position that may be corrected with an optimized data fusion algorithm.**

Keywords-vehicular simulation; sensors; communication; data fusion; prediction.

I. INTRODUCTION

Road safety have been considered primarily as a responsibility of each driver for a long time. Yet evolution of new vehicle technologies brought a renewal in this vision and allowed the development of new systems to assist drivers in their task: Advanced Driver Assistance Systems (ADAS) [1]. This possibility was granted thanks to the use of multiple on-board sensors and higher computational capabilities.

A similar trend was followed for road infrastructures and contributed to the development of new prevention techniques [2]. It is shown in [3] that these techniques correspond mainly to the analysis of vehicle and driver behavior and are based on detection, classification and tracking of different objects. These objects are evolving on a monitored location to prevent potentially dangerous situations and subsequently collisions and accidents. These new approaches oriented the choice of the different sensing systems to deploy on infrastructures, depending on the capabilities of the sensors and on the configuration on the chosen infrastructure itself. With more and more sensors involved to ensure better knowledge on the monitored area, data fusion has become a necessity. In [4], models, architecture, opportunities, and applications of data fusion among Intelligent Transport Systems (ITS) are described and show promising results.

Monitoring an intersection allows flexibility and the choice of sensors may be done by considering both the environmental factors and the particularities of the intersection. The concept introduced in this work offers a solution customizable to fit any road network and aims to perform with any type of sensors. In this regard, scenarios build in the chosen simulator are simple enough to accommodate all kind of sensors. Results presented in the review [5] outlined the good performances and relatively low-cost of camera-based solutions thanks to the progress of computer vision techniques. Although, their implementation requires most of the time the application of deep learning techniques. However, in order to keep the problem to a simpler state, other technologies are considered, such as: Radio Detection And Ranging (RADAR) or Light Detection And Ranging (LiDAR). Shirazi et al. in [3] give additional information on the trending sensors used for entity detection around intersections. While showing once again the potential of cameras, it also reveals comparable performances for LiDARs and rises only one main drawback: its high cost. However, this represents no obstacle for this project as simulations allow easy implementation without consideration of the cost. LiDAR will then be the chosen type of sensor at this stage.

Some more work has been exposed in [6], focused on the use of Dedicated Short-Range Communication (DSRC) and especially Vehicle-To-All (V2X) communications. It shows the potential of DSRC in intersection safety, as well as the implication of countries in its development. V2X demonstrates strong results while associated with the sensing equipment's that can be deployed in the infrastructure. Connected vehicles are considered as a highly potential step of the evolution of cars which may appear progressively, sooner than any automated driving vehicles from medium to high autonom level. Including communication technologies in any upcoming road safety solution should be a must.

Advances in infrastructure technologies have also led to improvements in traffic management through intersection cooperation [7], enabled by vehicular communications. Once again, some very promising solutions have been presented, leading to new thoughts on the future of signalization. However, while independently showing very good results, intersection cooperation and vehicular safety analysis have not gathered as much interest and very few works have been done in the preservation of analysis results. While powerful algorithms run on a single intersection to ensure its safety, once any mobile entity leaves this intersection, all gathered data is lost and next intersections will have to execute similar processes to infer to the same information for its self-use. While taking advantage simultaneously of infrastructure equipped with multiple sensors, DSRC and data fusion techniques, this paper introduces the idea of shared data through an example of intersection network. For this purpose, a testing environment has been simulated on the vehicular simulator SCANeR Studio from AVSimulation where different scenarios were run to gather data from dedicated sensors, then processed in external models synchronized with the simulator.

Section 2 explains the new concept presented in this document. Then, Section 3 gives more details about the mathematical models used. Section 4 introduces the system implemented and the corresponding scenarios put together to test the performances of the system. And eventually, a fifth section gives hints for improvements before the conclusion.

II. System and Data Sources

This section presents the overall idea of the described system along with the different data sources handled.

A. General Idea

The concept presented here is about data exchange between road infrastructures. While equipped with a set of sensors, each intersection can produce some knowledge about the vehicular situation at its surroundings. These data are valuable and require important resources to be generated but are also generally thrown away as soon as the detection is lost. Thus, the idea is to keep this awareness and to share it with other intersections in order to simulate sensor inputs for connected intersections. Based on this, a detection and prediction process are put together to anticipate the arrival of the travelling announced vehicle. In this paper, the performance of such a system are questioned and results are showed. The intention is to create an anonymous track of vehicles on the road, while alleviating the computed detection task of each intersection.

B. Data sources

Sensors chosen to gather data at intersection are DSRC and LiDARs. Both represent popular sensors adapted to the automotive industry. They can get data about all kind of vehicles with a stronger confidence for connected vehicles.

1) DSRC

DSRC is a vehicular communication protocol based on broadcasting messages, called Basic Safety Messages (BSM), to inform connected entities within the communication range about vehicles intentions [8]. It participates in the creation of an ad hoc network of vehicles, exchanging data in order to evolve in a secure environment. Among the information contained in every BSM, the position of the emitting vehicles, its current speed and acceleration are the ones that are the most interesting for this work. By gathering those data, it will be possible to keep an awareness of all connected vehicles present at the intersection.

An important note to make here is that this paper is not focused on disputing communication scheme. This point may be addressed in future works. That's why, at this point, all communications are assumed perfect.

2) LiDAR

LiDAR is an active light-sensitive sensor that allows detection of obstacles in a specific field of view. Its strength rests on its ability to give accurate information of obstacle detected up to hundreds of meters [9]. It has the advantage of providing distance data and performing by day or night. But it may be affected by extreme weather (heavy rains, fogs) and some reflective surfaces.

3) Intersection data

Intersection data corresponds to the addition we are putting forward in this paper. It corresponds to the shared information that each intersection will send to other adjacent intersections. These data are based on the prior output of any detection and classification algorithm achieved by other sensors: the LiDAR and DSRC in this case.

From its knowledge, any intersection will be able to determine leaving vehicles with their status: exit used, last speed and acceleration measured. It will then inform connected intersections of the arrival of a new vehicle. Upon reception of the intersection message, the target infrastructure is expected to build a continuous prediction model corresponding to the evolution of the anticipated vehicle until it appears in the field of detection of its own sensors (if it appears).

A big interest of this new element relies on the fact that it can convey any type of data for any detected object that the intersection is capable of detecting. With LiDARs and DSRC chosen, each intersection should be able to detect any vehicle passing by the intersection, with a stronger belief in detection of connected vehicles. Although, intersections can only send data about vehicles monitored: it will not be able to predict the apparition of a newcomer.

III. MATHEMATICAL MODELS

This section focuses on a description of the equations used in the processing of the data from each described sensor.

A. Cluster and single detection

1) LiDAR

Simulated LiDARs in SCANeR Studio 1.7 return a matrix of all distances measured by each programmed beam including: obstacles: vehicles as well as roads and signalization sign. For the environment set in this simulation and to minimize the data processing task, a 375-beam configuration has been chosen. It corresponds to a matrix of 15 rows and 75 columns, each beam separated by 3 degrees. This configuration ensures best processing performance but shows less accurate detections at long distance where beams are widely separate, increasing the detected position error. Two lidars, face to face, are monitoring each intersection as represented in Fig. 1. All measures returned are perfect.

In the context of a static LiDAR placed at a known position, detecting new object C_L in the environment can be made by taking a capture of the empty intersection C_{Ref} and comparing each new capture C_c with this reference (capture).

$$C_L = C_c - C_{Ref} \tag{1}$$

This method works well in a simulated environment with a few possibilities of noise due to uncontrolled elements appearing in the scenario. Thus, to reduce these possibilities, a filter is required. In this case, where this problem is not specifically approached, a simple two-bound threshold is applied to guarantee that all detected objects may correspond to vehicles. From the resulting matrix C_L, we can retrieve clusters of active beams corresponding to detected obstacles. The isolation of each of these clusters, in each column of a new matrix SC_L, allows the inference of listing spatial points forming each cluster. Hence, each cluster can be assimilated to a single center of gravity and its possible dimensions in final LiDAR detection matrix D_L.

Figure 1. LiDARs field of view.

$$Pos_i = \begin{bmatrix} x_i \\ y_i \\ z_i \end{bmatrix} = \begin{bmatrix} \bar{x}_{SC_L,i} \\ \bar{y}_{SC_L,i} \\ \bar{z}_{SC_L,i} \end{bmatrix} \tag{2}$$

for $i \in [0, \text{nb cluster}]$

$$Dim_i = \begin{bmatrix} S_{X_i} \\ S_{Y_i} \\ S_{Z_i} \end{bmatrix} = \begin{bmatrix} Max(x_{SC_L,i}) - X_i \\ Max(y_{SC_L,i}) - Y_i \\ Max(z_{SC_L,i}) - Z_i \end{bmatrix} \tag{3}$$

$$D_{L,i} = \begin{bmatrix} Pos_i \\ Dim_i \end{bmatrix} \tag{4}$$

Where $x_{SCL,i}$, $y_{SCL,i}$ and $z_{SCL,i}$ are the coordinates of each point belonging to cluster i, respectively.

2) DSRC

Data returned by the DSRC correspond to position, speed and acceleration of connected vehicles. From this, a simple noise μ corresponding to GPS error (set to five meters) is applied to the perfect position returned by the system [10].

$$D_{DSRC,j} = \begin{bmatrix} Pos_j \\ v_j \\ a_j \end{bmatrix} + \mu \tag{5}$$

for $j \in [0, \text{nb connected vehicle}]$

$$\text{with } Pos_j = \begin{bmatrix} x_j \\ y_j \\ z_j \end{bmatrix} \text{ and } \mu = \aleph(0, P) \tag{6}$$

where Pos_j corresponds to the coordinates of the connected vehicle j, v_j to its speed and a_j to its acceleration. Lastly, μ represents a white Gaussian noise with a covariance P corresponding to the GPS location error.

For synchronization purpose, BSMs are supposed to be all send at the same time and queued by the nearest intersections in communication range. Target intersection will then be able to build the current vehicle situation according to this input.

3) Intersection data

Intersection data depend on the result of the data fusion of all providing sources. Once this data fusion gives the current result, obstacles are located within the intersection and if some of them are flagged as leaving at a specific exit, then last fused data F_l are sent to the corresponding intersection.

$$F_l = \begin{bmatrix} Pos_l \\ v_l \\ a_l \end{bmatrix} \quad \text{for } l \in [0, \text{nb leaving object}] \tag{7}$$

Where Pos_l correspond to the fused coordinate of leaving object l, v_l to its last registered speed and a_l to its last registered acceleration.

The broadcasted data corresponds to the set of objects leaving the intersection at the specific exit. It mainly contains the moment of exit and, if available, the speed and acceleration of the object. This last data may not be accessible in the case of a non-connected vehicle, as no tracking is implemented for objects detected by LiDARs.

Upon reception of any intersection data, the target intersection is expected to resort to a prediction model of the listed vehicles possibly incoming. At this stage, each intersection knows the position of each connected intersection, the configuration of the road section in between (topology and speed limit) and the last position and possible speed and acceleration of the departing object are known by the target intersection.

For this simple scenario, the prediction part of a simple Kalman Filter is applied. In order to do so, the linearization of the vehicle dynamic is done by projecting the three Cartesian position coordinates to a single straight road between both involved intersections. In the case of this particular scenario topology, reducing the location to a one-dimensional problem (represented by variable d_k) is made by a compelling projection corresponding to the Manhattan distance (8) between the two involved intersections (with target intersection as origin).

$$d_0 = proj(Pos_l) \tag{8}$$

This projection gives a rather good estimate of the position of the vehicle during time but rests on the need of a mapping table. Then, the prediction part of the Kalman filter is applied with all known data.

$$\begin{cases} d_k = d_{k-1} + \Delta t \times v_{k-1} + \dfrac{\Delta t^2}{2} \times a_k \\ v_k = v_{k-1} + \Delta t \times a_k \end{cases} \tag{9}$$

Where d_k corresponds to the distance remaining to the next intersection (based on the projection of the coordinate), v_k corresponds to the predicted speed of the vehicle and a_k to its acceleration. With the state vector being $X_k = [d_k \quad v_k]^T$ with $X_0 = [d_0 \quad v_l]^T$, we deduce the model, as follows:

$$X_k = F_k \times X_{k-1} + B_k \times U_k \tag{10}$$

with $F_k = \begin{bmatrix} 1 & \Delta t \\ 0 & 1 \end{bmatrix}$ as the state transition matrix and the control part corresponding to the acceleration influence and defined as follows:

$$U_k = \begin{cases} 0 & \text{if } v_{k-1} = \text{speed limit} \\ a_l & \text{otherwise} \end{cases} \tag{11}$$

$$B_k = \begin{bmatrix} \dfrac{\Delta t^2}{2} \\ \Delta t \end{bmatrix} \tag{12}$$

In (11), the acceleration is assumed to be constant until speed limit is reached. This acceleration corresponds to the last acceleration measured for the leaving vehicle a_l.

The prediction part of the covariance matrix P_k according to the Kalman filter is then defined as:

$$P_k = F_k \times P_{k-1} \times F_k + Q_k \tag{13}$$

$$\text{with } P_0 = 10^{-6}.diag(2) \tag{14}$$

$$\text{and } Qk = \begin{bmatrix} \Delta t^2 & \dfrac{\Delta t^3}{2} \\ \dfrac{\Delta t^3}{2} & \dfrac{\Delta t^4}{4} \end{bmatrix} \tag{15}$$

Where Q_k corresponds to the covariance matrix of the acceleration influence. Such that the final prediction model corresponds to:

$$\begin{cases} X_k = F_k \times X_{k-1} + B_k \times U_k \\ P_k = F_k \times P_{k-1} \times F_k + Q_k \end{cases} \quad (16)$$

For every returned state vector X_k, it is possible to infer to the corresponding d_k to the actual Cartesian position thanks to the projection table.

B. *Data fusion*

Data fusion is applied on three nodes, as in Fig. 2:

- Between data from both LiDARs
- Between data from LiDAR and intersection data (IT Data)
- Between previous fused data and DSRC

All three of these data fusion correspond to a data association processed with a method inspired from the covariance intersection method [11]. The principle is to get both the set of position and covariance matrix measured or predicted in detection methods above and trying to find the best match between each sensor to fuse. Covariance Intersection method states that:

$$P_{pf}^{-1} = w_1 \times P_{P_1}^{-1} + w_2 \times P_{P_2}^{-1} \quad (17)$$

$$Pos_f = P_{pf} \times (w_1 \times P_{P_1}^{-1} \times Pos_1 + w_2 \times P_{P_2}^{-1} \times Pos_2) \quad (18)$$

Where Pos_i corresponds to the Cartesian position of considered points and P_{P_i} corresponds to the covariance matrix associated to Pos_i. All position measures are made relative to the center of the intersection processing the vehicle detected. The optimization problem, where w_k must be computed, is not of interest in this situation. More simply, this same variable is estimated (19) with each covariance matrix from sensors to fuse (P_{P_1} and P_{P_2}) and then the fused position (Pos_f) and covariance matrix (P_{pf}) are determined and compared to both initial positions (Pos_1 and Pos_2) to

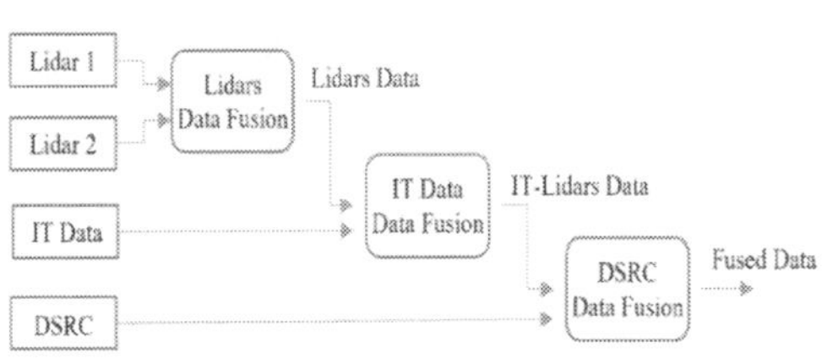

Figure 2. Data Fusion Scheme.

assess possibility of matching.

$$\frac{w_1}{w_2} = \frac{m_1}{m_2} = \alpha \quad (19)$$

$$\text{with } m_k = Max(P_k(i,j))$$

$$\text{And } w_1 + w_2 = 1 \text{ so } w_2 = \frac{1}{1+\alpha} \quad (20)$$

$$\begin{cases} dist(Pos_f, Pos_1) < m_1 \\ dist(Pos_f, Pos_2) < m_2 \end{cases} \Rightarrow S = \{Pos_f\} \quad (21)$$

Where $dist(Pos_f, Pos_k)$ corresponds to the Euclidean distance between the two position vectors and S corresponds to the set of all fused position vector verifying the condition in (19).

An area estimation of each matching possibility ($A(S(l))$) is then calculated and the wider area is retained as best match (BM).

$$BM = S(k) = Pos_{f,k} \quad (22)$$

$$\text{where } A(S(k)) = Max(A(S)) \quad (23)$$

$$\text{and } A(S(l)) = \pi \times \prod P_{f,l}(i,i) \quad (24)$$

It is to be noted that each captured data frame is considered independent of the others. For this purpose, no tracking was implemented during active detection of mobile entities, except when the detected vehicle was considered as leaving the intersection. Region-based comparisons grant this distinction and trigger a tracking process where a Kalman Filter is used to follow leaving vehicles.

IV. EXPERIMENT AND RESULTS

This section depicts the configuration of the scenarios built to tests this system, the experiments conducted, and the results obtained.

A. *System presentation*

For the purpose of this article, a simulated environment has been built using the vehicular simulator SCANeR Studio from AVSimulation. This environment consists of a succession of three intelligent intersections highlighted in Fig. 3. While all these intelligent intersections are equipped with sensors, we focus the study on the intersection in the middle. The other intelligent intersections are assumed to host perfect sensors and only their detection outputs are sent to the monitored intersection.

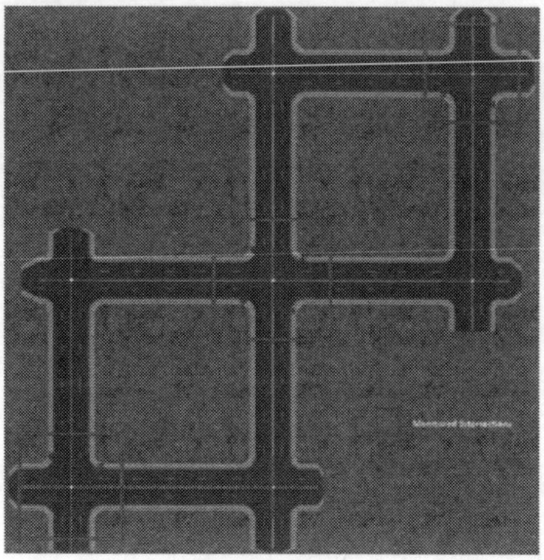

Figure 3. Simulated environment with monitored intersections highlighted.

This simulated topography is based on the modern representation of urban road network with a Manhattan grid network. It also allows easier distance calculation when needed. However, the vehicle flow is limited to few vehicles compared to real urban scenarios where hundreds of vehicles may be present. This scenario will be used as an early stage of the concept introduced.

All three of these intersections are identical and composed of four pairs of entries/exists following two axes, as shown in Fig. 4: north-south and east-west. Each of these entries is marked by a stop sign and contains two lanes: left lane allows only left turn while right lane allows all remaining other directions.

The scenario also contains four vehicles of different dimensions driving within the intersections and following a specific pre-defined path assigned to each of them.

For the processing part of each intersection, control models have been implemented externally to compute all exchanged data. Simulated environment and control models communicate through UDP and are synchronized so that reaction models depend on the vehicular simulator outputs. In this regard, these models are configured to ran at a faster rate than the simulator to efficiently process all data in due time.

The physical platform is currently only composed of a Windows 10 computer running SCANeR Studio version 1.7. But this system has been thought so that it can be implemented on a real-time platform to perform Software-In-The-Loop (SIL) based validation. Indeed, data exchange through UDP allows the communication with any external devices such as, for example, a real-time platform from Opal-RT Technologies. This would allow easy handling of any synchronization matter within the platform and ensures the access to greater computing resources for more complex systems.

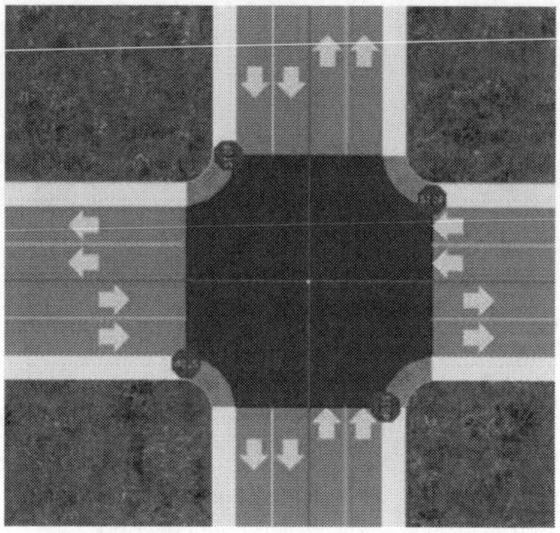

Figure 4. DSRC Entries and exits of each intersection.

B. *Experiments details*

Two main experiments will be realized to attest the performances of this system. First, the contributions of the new system will be observed through the comparison of the result of the data fusion with IT Data and with the initial output of the lidars data fusion.

Second, as the main sensors are LiDARs and DSRC, the number of connected vehicles may have some effect on the performance of the system. For that reason, four scenarios are considered in Table 1.

These scenarios can relate to the evolution of the vehicle population with time (DSRC penetration ratio), following the progress of the automotive industry. And, as a matter of fact, may show the relevance of the proposed system.

Performance factors will correspond to the influence of the intersection data on the detection of vehicles entering the new intersection. For this analysis, comparison factors correspond to:

- Position error of the detected vehicle compared to its true position;
- Data covariance ratio between results of data fusion with and without Intersection Data.

TABLE I. LIST OF SCENARIOS

#	Scenarios
A	Only connected vehicles
B	Number of connected vehicles greater than 50%
C	50% of connected vehicles
D	Number of connected vehicles lesser than 50%

Three minutes long scenarios corresponding to the ones described in Table 1, are monitored. Each of these scenarios presents eleven intersection crosses by vehicles from the simulation, nine of them announced by other intersections. Details about detections made in main intersection are registered at each time and saved for processing, for a total of 12 GB of data. Next section presents a condensed study of this data, depending on the comparison factors chosen.

C. Results and discussions

From the first experiment, two graphics are generated. Fig. 5 presents the distance error between the real position of the vehicle and the detected position. Values from the result of IT-LiDARs data fusion are compared with LiDARs fusion. This measure is limited to each time that a fusion is operated with an IT Data and then occurred only when a vehicle from another intersection is coming in range of the target intersection.

Fig. 5 shows that the fusion with IT Data tends to move detections away from their true position with some extreme cases as for the third successful fusion attempt where the resulting fusion comes from two separated vehicles detected by each sensors with a covariance large enough for the data fusion algorithm to consider it possible. However, the average distance error of the IT-LiDARs fusion results (counting the wrong fusion) remains around ten meters, still corresponding to a standard GPS error.

On the other hand, Fig. 6 shows the ratio of covariance for the same detections as before, comparing results from the LiDARs fusion and results from IT-LiDARs fusion. It shows that in second case, the covariance is reduced in average by approximatively thirty percent. This implies a better precision of the detection even with a bigger distance error to the real position. The fusion error specified earlier can also be observed in this new graphic where the covariance ratio exceed one in value. Furthermore, another exceeding value can be observed which also corresponds to a fusion error. This gives hint of improvement for the algorithm.

The second experiment focuses more on the study of the impact on the system of communicating vehicles. From it, we obtain both boxplots diagram in Fig. 7 and 8. In this case, all detections are considered, and pertinent data are summarized in presented graphs.

Fig. 7 presents the distance errors between detected and real position of vehicles for each scenario described in Table 1. And Fig. 8 details the covariance ratio between the scenarios with IT Data and the ones without it. Results obtained allow to confirm the ones obtained in the previous experiment: slight distance shift of the detection position and better covariance ratio. Also, what is more interesting to notice here is the relative absence of changes between the

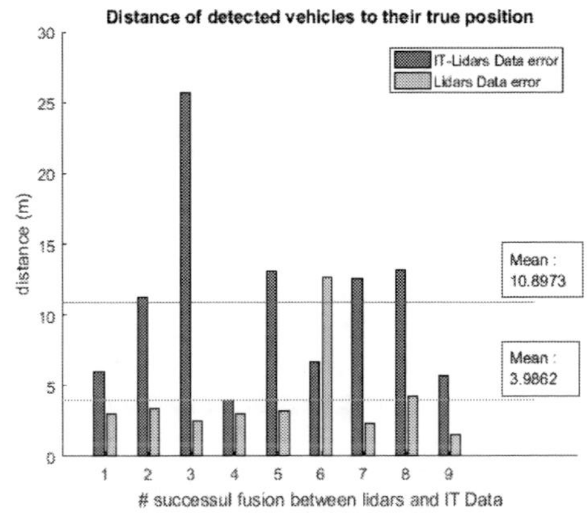

Figure 5. Distance error between detections and real position.

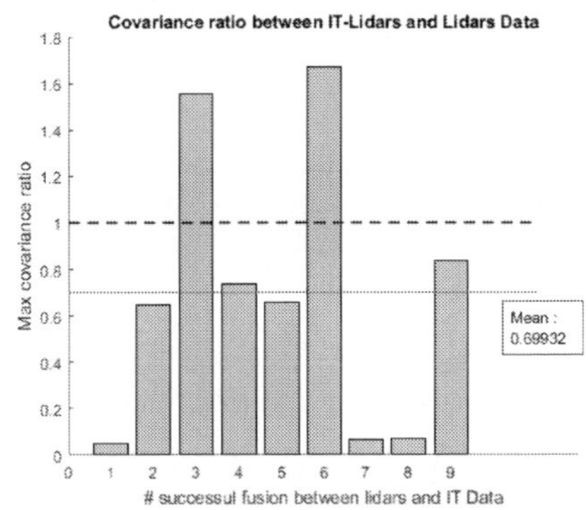

Figure 6. Covariance ratio between IT-LiDARs and LiDARs Data.

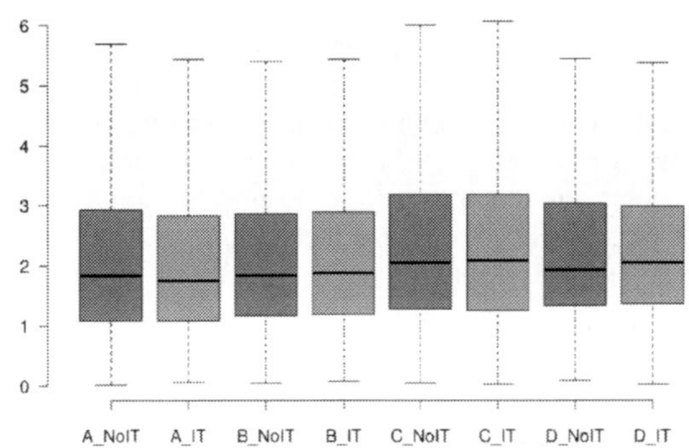

Figure 7. Distance errors between detected and real vehicles for each described scenarios – with and without Intersection Data.

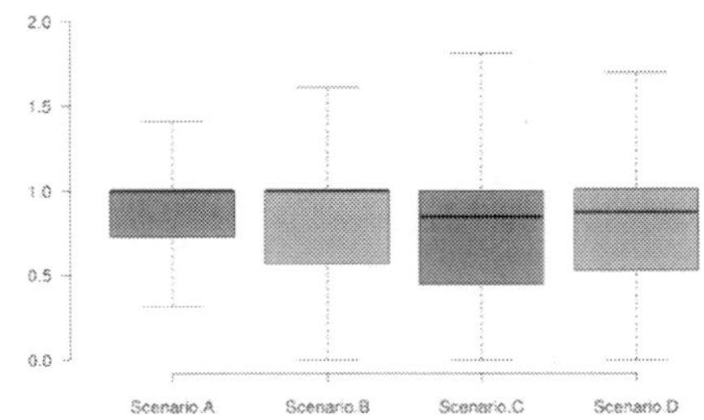

Figure 8. Covariance ratio between results with Intersection Data and without for each scenarios.

different scenarios. From this observation, we can conclude that the system is quite robust against the variation of communicating vehicles and allow to perform a relatively more precise detection of all vehicles in the vicinity of the intersection.

The scenario presented for this system is simple and does not take into consideration all variations that may happen in most real-life scenario: vehicles driving through the road network may leave at any moment without any warnings, stop in a place between two intersections or even face any sort of possible obstacles that may slow down its progression. All those possible outcomes will have an impact on the system, resulting in the loss of the data gathered about this vehicle or its mix with other data from another vehicle circulating on the same road sections. The first case does not represent much trouble as following detection of the vehicle will only be considered as new. However, the second case might lead to some data mismatch which may affect the detected behavior of vehicles on some road sections between intersections. The comparison of the results obtained in these simulations with the data from real-life scenarios is of great interest and will be the topic of a future work. It is to be noted that some adjustments may be needed to guarantee that the vehicular simulator can handle further vehicle loads in order to match more realistic urban scenarios.

V. CONCLUSION AND FUTURE WORK

Among all the assumptions made in this work, communications are considered perfect which surely affect positively the results presented. A more realistic simulation should involve more realistic communication models. That's why it would be necessary to switch to a later version of the vehicular simulator used or even add an intermediary software specialized in communications and include DSRC to enhance the current model.

Detection algorithms used in this work lacks optimization. The covariance intersection method implemented searches only for a possible fusion result and elect the best choice based on the largest common covariance area between two detections. Better adjustments could be made to enhance the potential of this data fusion and reduce the covariance factors of detections.

On a similar thought, the Kalman prediction performed to follow vehicles between intersections is also open to better performances. Some pre-study of typical vehicle dynamic evolution could be used to set a speed profile for better anticipation. This could contribute to real improvement of the results gathered in this work and lead to an acceleration of the detection.

Also, other type of sensors and data fusion methods will be applied to guarantee good results of the introduced concept. This analysis will be introduced as soon as scenarios with heavier vehicle load will be implemented for closer fidelity to urban scenario.

To conclude, the work presented in this article introduces a different handling of the data gathered by intelligent road infrastructures in urban area. The recycling of old detection data towards the whole intersection network was presented as a mean of preparation for smart cities to welcome intelligent and automated driving vehicles among other vehicles. Results presented have demonstrated that the system is effective with no regards to the number of communicating vehicles present. It still needs to be improved but already has some interesting outputs when applied to the discussed scenarios.

REFERENCES

[1] M. Lu, K. Wevers and 'R. Van Der Heijden', "Technical Feasibility of Advanced Driver Assistance Systems (ADAS) for Road Traffic Safety", Elsevier, Transportation Planning and Technology, 2005.

[2] A. Shalom Hakkert and V. Gitelman, "Thinking about the history of road safety research: Past achievements and future challenges", Elsevier, Transportation Research Part F: Traffic Psychology and Behaviour, 2014.

[3] M. S. Shirazi and B. T. Morris, "Looking at Intersections: A Survey of Intersection Monitoring, Behavior and Safety Analysis of Recent Studies", in IEEE Transactions on Intelligent Transportation Systems, vol. 18, no. 1, pp. 4-24, January 2017.

[4] N.-E. El Faouzi and L. A. Klein, "Data Fusion for ITS: Techniques and Research Needs", Elsevier, Transportation Research Procedia, 2016.

[5] Zi Yang, Lilian S.C. Pun-Cheng, "Vehicle detection in intelligent transportation systems and its applications under varying environments: A review", Elsevier, Image and Vision Computing, 2018.

[6] L. Le, A. Festag, R. Baldessari and 'W. Zhang', "Vehicular wireless short-range communication for improving intersection safety", in IEEE Communications Magazine, vol. 47, no. 11, pp. 104-110, November 2009.

[7] L. Chen and C. Englund, "Cooperative Intersection Management: A Survey", in IEEE Transactions on Intelligent Transportation Systems, vol. 17, no. 2, pp. 570-586, February. 2016.

[8] J. B. Kenney, "Dedicated Short-Range Communications (DSRC) Standards in the United States", in Proceedings of the IEEE, vol. 99, no. 7, pp. 1162-1182, July 2011.

[9] E. D. R. Shearman, E. G. Hoare and 'A. Hutton', "Trials of automotive radar and lidar performance in road spray," IEE Colloquium on Automotive Radar and Navigation Techniques (Ref. No. 1998/230), London, UK, pp. 10/1-10/7, 1998.

[10] S. Hong, M. H. Lee, H.-H. Chun, S.-H. Kwon and 'J. L. Speyer', "Observability of error States in GPS/INS integration", in IEEE Transactions on Vehicular Technology, vol. 54, no. 2, pp. 731-743, March 2005.

[11] A. Poularikas, M.E. Liggins, D. L. Hall, 'J. Llinas' "Distributed Fusion Architectures, Algorithms, and Performance within a Network-Centric Architecture", from Martin E. LigginsDavid L. HallJames Llinas," Handbook of Multi-sensor Data Fusion, Theory and Practice." second Edition, CRC Press Taylor & Francis Group, Boca Raton London New York, PP: 411-43, 2009.

Car-driving Interface with Load Cells for Upper-extremity-disabled People

Yoshitoshi Murata
Faculty of Software and Information Science
Iwate Prefectural University
Takizawa, Japan
e-mail: y-murata@iwate-pu.ac.jp

Yuto Higuchi, Takaya Abe
Radio Network Division
DOCOMO Technology Inc.
Tokyo, Japan
e-mail: {yuuto.higuchi.nb, takaya.abe.vb}@nttdocomo.com

Abstract— **Disabled people generally want to stand on their own two feet, and achieving mobility is an important step in satisfying that desire. A steering-operation unit for disabled people with disability in their arms was developed and experimentally evaluated. The unit consists of a set of load-cell sensors, one for turning right and one for turning left. The driver steps the right or left load cell to turn the car right or left. The magnitude of the driver's stepping force is converted to a voltage and input to the power-steering motor. The angular velocity of the steering wheel corresponds to that voltage. As a result of this configuration, the driver can drive a car just by moving their foot and intuitively selects the load-cell they must apply by foot to turn the car. Experimental results using a standard car fitted with the developed steering operation unit show that disabled people can drive the car with their foot in a manner close to that achieved with a steering wheel.**

Keywords-Car driving interface; disabled people; load cell; steering operation; foot operation.

I. INTRODUCTION

People who have physical disabilities generally want to stand on their own two feet, and achieving mobility on their own is an important step in satisfying that desire. One way for them to enhance mobility is by driving cars fitted with driving-assistance devices. However, only a few driving interfaces that enable disabled people, especially people with arm and wrist disabilities, to drive cars have been developed.

The first auxiliary device for people with arm and wrist disabilities, the original of Honda's Franz system [1], was developed in the 1960s. As for this system, the car is operated with the feet only. Since the steering wheel is turned by pumping the pedals, its operation is not intuitive.

The steering wheel in the system developed by Wada and Kameda in 2009 is controlled with a joystick, and the brake and accelerator are controlled with another joystick [2]. This system has aided many disabled people, but a certain amount of arm strength is needed to operate the joysticks. Moreover, the levers onto which the joysticks are fixed have to be customized to match the hand positions of individual users. In any case, mechanical devices, such as these lack flexibility and must be customized for individual users; hence, cars customized is this manner are inherently expensive.

At a glance, the autonomous car would be ideal solutions for disabled persons. In addition to those developed by Google, autonomous cars have been developed by many car manufacturers [3]. However, many disabled persons would like to not only reach a destination but also enjoy driving the car. Although several technologies, such as automatic braking and lane keeping, developed for autonomous vehicles would enable them to drive safely, riding in an autonomous vehicle would not satisfy their desire to drive.

A current realistic solution for disabled persons is to control steering with bodily appendages they can move. We have developed several kinds of steering-operation units until now. In case of the first operation unit we developed, namely, an angle sensor, the angle of the steering wheel changes according to the angle of a joint of the driver, such as the wrist or ankle [4]. As for the result of an experiment using the operation unit fitted in an experimental electric vehicle, the driver could drive the vehicle with the operation unit and achieved steering close to that with a steering wheel on gently curved roads. However, it was difficult for the driver to turn the car at intersections. To address that problem, we changed the angle sensor to a load-cell sensor [5].

We fitted the load-cell sensor in a standard car with a power-steering unit. A steering-control unit was designed for people with arm and wrist disabilities, and it was operated by the left foot. Since we noticed that it was difficult for a disabled driver to drive a car by forward and backward movement, not right and left movement, the steering-control unit consisted of a set of right and left load-cell sensors. The driver stepped on the right or left load cell to turn the car right or left. Results of an experiment using the developed steering-operation unit showed that the driver could drive the car with their left foot in a manner close to that achieved with a steering wheel on roads with not only gentle curves but also sharp bends.

After introducing related works in Section II, actions by body parts suited for driving are explained in Section III, and results of our previous experiments using the angle-sensor operation unit fitted in an experimental vehicle are presented in Section IV. As for our latest study, described in Section V, we investigated whether the load-cell-sensor operation unit can be applied as a driving-interface unit. Section VI concludes this paper.

II. RELATED WORKS

Highly functional steering wheels have been developed by researchers in car manufactures or universities to easily drive a car for usual healthy people [6][7]. Sucu and Folmer challenged to develop a steering wheel for blind drivers [8]. Since the purpose of this study is to design an operation interface mainly for disabled people who has disabilities in the arm and/or wrist, we introduced existing advanced driving interfaces for people who have difficulty moving their arms and/or hands.

A. Franz system

The Franz system used by Honda is aimed at people who have difficulty moving their arms and hands [1]. The user operates a car fitted with the system with only their feet. It was originally implemented in a Honda Civic in 1982, which was the first vehicle to introduce the Franz system in Japan. It has now been implemented in a Honda Fit. The steering wheel is turned right or left by pumping a steering pedal (see Figure 1). The transmission is shifted into drive by lifting the selection bar, into reverse by pushing it down, and into park by pushing it further down. The turn signals and windshield wipers are operated by turning levers with the right and left knees. Power windows and lights are controlled by flipping switches up or down with the right foot or knee.

(1) Steering pedal
(2) Steering box
(3) Brake lock button
(4) Selection bar for feet
(5) Side brake for knee
(6) Sub-brake for exercise

Figure 1. Honda's Franz system

B. Joystick operation

Several kinds of the joystick-operation unit are available. Such a unit is usually adopted for people who do not have enough strength to control a steering wheel, an accelerator pedal, or a brake pedal because of a disability, such as a spinal-cord injury. They use joysticks instead of a steering wheel and pedals. The operation unit showed in Figure 2 was developed by Wada and Kameda [2]. It is available as two types. One is a single stick, by which steering, braking, and acceleration are controlled with one joystick. The other is a double stick, by which the joystick on the right controls the steering and that on the left controls acceleration and braking. The double-stick type is shown in Figure 2.

The relationship between the angle of the steering wheel and the angle of the joystick is a polyline, as shown in Figure 3. It means that the driver can sensitively control the steering wheel around a neutral position and turn the wheel quickly when making a wide turn. People who can freely move their hands can drive cars with this device. However, such mechanical devices must be customized to fit individual users' disabilities and physical form.

Figure 2. Wada and Kameda's joystick driving interface

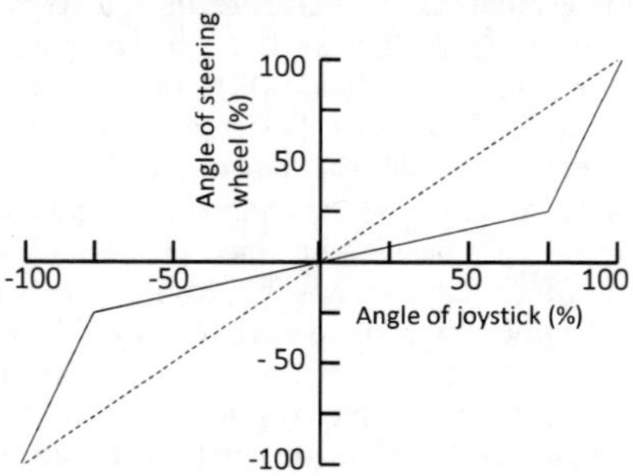

Figure 3. Relationship between angle of joystick and angle of steering wheel

III. DRIVING ACTION BY BODY PARTS

The purpose of this study is to design an operation interface for disabled people that is operated by body parts moving smoothly instead of by hand. In a previous study, therefore, we measured car-control characteristics for several actions: rolling the ankle, moving the forefinger, moving the wrist, rolling the lower arm, moving the lower arm backward and forward, and moving the upper arm backward and forward [9]. The results obtained from the questionnaires (answered by 29 participants) are summarized in Table I. Most people chose the same action, such as rolling the lower arm, for motions that led to right or left movement of the car. For example, someone may move their finger down to turn a car to the right, while another may move their finger up to turn to the right. Hence, we obtained information about different motions by individuals by administering questionnaires before measuring the car-control characteristics.

However, the number of people who chose alternative actions, such as moving their fingers up or down, was roughly the same for actions that did not lead to right or left movement of the car. For example, 86% of people chose rolling their right lower arm to the right to turn the car to the right. However, 52% of participants chose "up" and 48% of them chose "down" for moving their left finger up or down.

The results listed in Table 1 show that the operation unit must be operated by the actions that lead to right and left movement of the car.

TABLE I. RESULTS FROM QUESTIONNAIRES ABOUT ACTIONS FOR TURNING THE CAR TO THE RIGHT

Left hand		Turn to right	Right hand		Turn to right
Finger	Up	15	Finger	Up	18
	Down	14		Down	11
Finger	Right	26	Finger	Forward (Right)	27
	Forward (Left)	3		Left	2
Wrist	Up	16	Wrist	Up	16
	Down	13		Down	13
Wrist	Right	24	Wrist	Right	27
	Left	5		Left	2
Lower arm	Forward	11	Lower arm	Forward	21
	Backward	18		Backward	8
Lower arm	Right	25	Lower arm	Right	28
	Left	4		Left	1
Upper arm	Forward	17	Upper arm	Forward	14
	Backward	12		Backward	15

IV. OPERATION UNIT USING ANGLE SENSOR

An operation unit using an angle sensor was developed and fitted in an electric vehicle. Three angle sensors were evaluated: a chromium-nickel sensor developed by the Research Institute for Electromagnetic Materials [10] and TMI-160 developed by Toyo Sokki Co., Ltd [11]. According to the results of the evaluation, TMI-160 was selected and attached to the parts of a subject's body, as shown in Figure 4. In this experiment, an experimental electric vehicle, PIUS [12], developed by MODI Co., Ltd., showed in Figure 5, was used. The driver shown in this figure has a disability due to taking the drug Thalidomide. Her arms are too short to operate a steering wheel. Nevertheless, she can operate a stick steering control system mounted on the vehicle chassis with her fingers. Servo motors were attached to the steering wheel, accelerator pedal, and brake pedal so they could be controlled by the sensor operation unit.

The drivability of the driving operation unit using the angle sensor was experimentally evaluated. The test courses used in the evaluation are a straight course, a square course, and a circuit course (see Figure 6). The drivability in terms of "straightness" and "turnability" were evaluated with the straight course and square course.

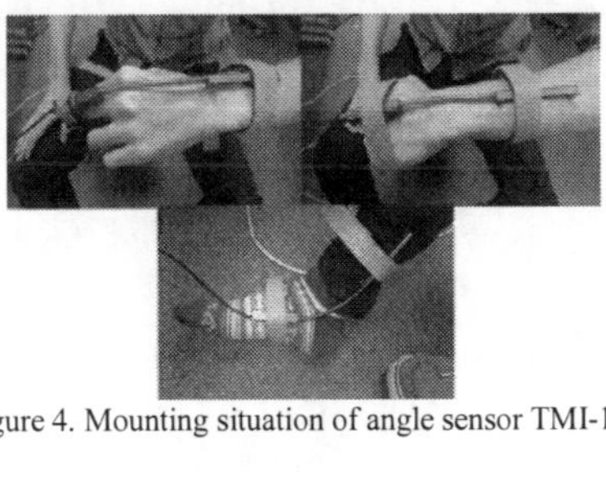

Figure 4. Mounting situation of angle sensor TMI-160

Figure 5. Experimental electric vehicle PIUS

Overall performance of the driving operation unit was evaluated with the circuit course.

In each test, the length between the right front tire and the center line on the straight and circuit courses was measured. Its average and Standard Deviation (SD) are listed in Table II. The standard deviation was assumed to shows the fluctuation. The number of participants in the tests was five. Sensors were attached to the middle finger, wrist, and ankle of the participants.

We measured the same data by the steering wheel operation to compare with them by the angle sensor unit. As introduced in Section III, it is difficult for a driver to imagine turning the car to the right or left from the up-down action of the thumb. Therefore, we suggested to the participants to turn their thumb upward. The action of the thumb and ankle control the right or left motion of the car in this position. In the case of driving by moving the ankle, we suggested to the subjects to tilt their knee outwards to easily image the right or left movement of their ankle. The SD of the measurements taken by the angle-sensor operation unit was greater than that of those taken by steering wheel. However, the difference in SDs was not significant.

TABLE II. DRIVING CHARACTERISTICS IN THE CASE OF DRIVING ON THE STRAIGHT AND CIRCULAR COURSES USING THE ANGLE-SENSOR OPERATION UNIT

Course	Operation	Running time [m.:s.]	Average [cm]	SD [cm]
Straight	Middle -finger	0: 58	53.3	6.6
	Wrist	1: 03	56.2	7.0
	Ankle	1: 04	55.8	6.2*
	S. wheel	0.59	54.0	6.1
Circuit	Middle -finger	6: 12	56.3	8.2
	Wrist	6: 22	56.5	10.0
	Ankle	7: 03	53.3	9.4
	S. wheel	6: 17	55.3	7.0

* One participant drove out of course; therefore, their data was discarded.

TABLE III. EVALUATION EXPERIMENTS USING A SUBJECT WITH DISABILITY

Device / BP	Running time [m.:s.]	Length from center line Average [cm]	SD [cm]
Stick operation unit	8:37	61.3	8.7
Ankle	8:22	59.0	5.8

*BP: Body parts

TABLE IV. THE NUMBER OF TIMES THE DRIVER DROVE OUT OF COURSE AT THE CORNERS

	Device/BP	Clockwise A	B	C	Av	Counter clockwise A	B	C	Av
The number of course out	SW	0	0	0	0	2	1	1	1
	Mid-finger	7	4	6	6	8	3	4	5
	Wrist	9	7	5	7	9	6	3	6
	Ankle	12	9	10	10	11	8	9	9
Rate of course out	SW	0	0	0	0	17	8	8	11
	Mid-finger	58	33	50	47	67	25	33	42
	Wrist	75	58	42	58	75	50	25	50
	Ankle	100	75	83	86	92	67	75	78

*BP: Body parts, A/B/C: Participants

A research partner, Ms. Masuyama was requested to evaluate the driving operation unit. She usually drives a car fitted with the Frantz system. She operated a stick operation unit by her fingers showed in Figure 5, and a sensor attached to her ankle. She drove the car fitted in this manner on the circuit course. The results of the measurements taken by the angle sensor are listed in Table III and plotted in Figure 7. According to these results, she could drive the car very well. Especially, her fluctuation characteristics were better than those in the case of using the steering wheel.

Drivability in terms of turning was also evaluated by using the square course (Figure 8). Three participants drove three laps on the square course in both the clockwise and counter-clockwise directions. The driver turned twelve corners in each direction. In this experiment, the number of times the driver drove out of course at the corners was recorded. The results of the experiment are listed in Table 4. According to these results, unfortunately, every participant could not turn the corners well by using the angle-sensor operation unit.

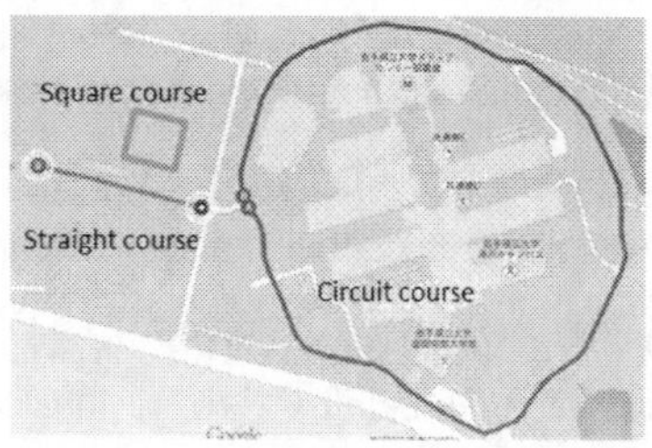

Figure 6. Test courses

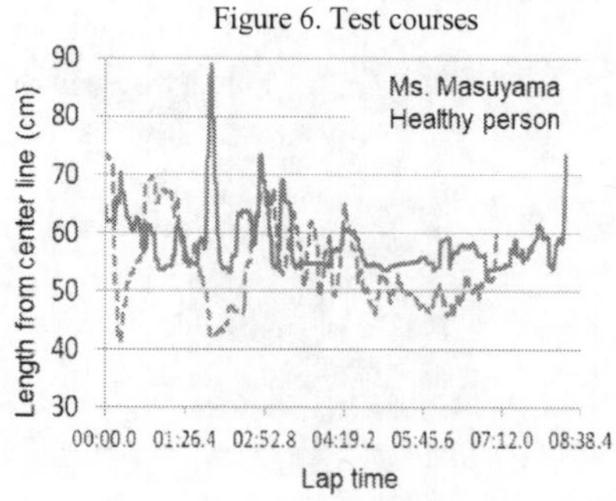

Figure 7. Distance from center line

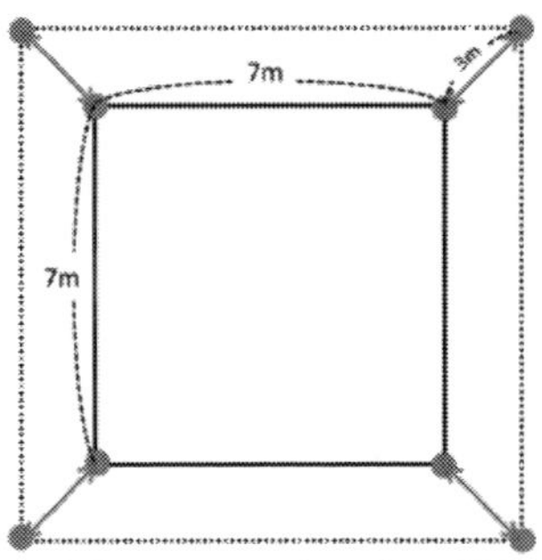

Figure 8. Square course for right and left turn

V. Operation Unit Using Load Cell Sensor

Participants could smoothly drive the car with the angle-sensor operation unit on the circuit and straight course, but they could not drive so well on the square course. Lock-to-lock angle of the steering wheel is about 800 degrees. On the other hand, the maximum range of movement of their body parts is roughly 90 degrees. That means it is very difficult for the angle-sensor operation unit attached to the body part to correspond to the lock-to-lock angle of the steering wheel. In consideration of that fact, the angle sensor was replaced with a load cell sensor. First, whether a load cell sensor operation unit can control the car with the hand in the same manner as the load cell operation unit was experimentally examined. According to the results of that experiment, the driver could drive the car by applying enough force. Based on that result, a load cell operation unit operated by foot was developed.

The load cell sensor operation unit is available for use as a driving operation unit. Hence, we designed a driving operation unit for people with arm and wrist disabilities. Since this operation unit comprised right and left load cells (see Figure 9), the driver can intuitively select a load cell and step on it to turn the car to the right or left. In this experiment, a load cell, Pedal Force Sensor (PFS) developed by Toyo Sokki co. ltd [5] was selected because it is easy to step on. The operation unit was connected to the power-steering motor of a standard car (namely, a Nissan Micra) as the test car. The operation unit was mounted on the car as shown in Figure 10. In the case of the power-steering motor, the steering rotation speed is changed according to an impressed voltage. Therefore, intensity of force applied to the load cell corresponds to the rotation speed of steering wheel (not to the angle of rotation of the steering wheel). Two control schemes were examined to determine the relationship between intensity of force applied to the load cell and control power supplied to the steering unit. Each control scheme is listed as follows:

"Digitize" scheme

(1) No intensity: the steering wheel is free. It returns to the neutral position by the force from the tires.
(2) Very weak: a foot is just put on the load cell, no stepping; the steering wheel stops at its position.
(3) Weakly stepping on: the steering wheel rotates slowly.
(4) Strongly stepping on: the steering wheel rotates quickly.

"Continuous" scheme

(1) Same as digitize scheme (1).
(2) Same as digitize scheme (2)
(3) Stepping on: rotation speed of the steering wheel changes according to intensity of the stepping force.

Drivability of the car fitted with the driving operation unit using the load-cell sensor operated by foot was experimentally evaluated. The results are shown in Table V. The test courses used in the evaluation were the straight course and circuit course (described in Section IV). Measurement data were also the same as that presented in Sections IV and V. The number of participants was seven. The driving characteristics are a little worse than those of the car operated by the angle-sensor operation unit and load-cell-sensor operation unit for the hand.

After that, turning drivability was evaluated. Since drivability of the car driven by the load cell sensor operation unit for the foot is superior to that in the case of the angle-sensor operation unit, the right/left-angle corner was used, and a white line was painted on the course to indicate traffic lanes, as shown in Figure 11. Each of the five participants in the test drove the car three times by using the load cell operation unit for the foot in each direction. For the sake of comparison, they also drove the car with the steering wheel. In this experiment, the number of drivers went out of the traffic lane was counted. According to the results of the experiment, every participant did not drive out of traffic lane in both the steering wheel and load cell operation unit for foot.

Since the drivers could drive the car well with a combination the digitized scheme and the power-steering unit, it was concluded that the drivers can also drive a standard car fitted with an operation device that has a neutral step and three steps each for right and left turning, totally seven steps. As such a device, an angle or load-cell sensor are usable, but not necessary.

TABLE V. DRIVING CHARACTERISTICS IN CASE OF STRAIGHT AND CIRCULAR COURSES BY OPERATING THE LOAD-CELL-SENSOR OPERATION UNIT BY FOOT

Corse	Operation Method	Length from center line SD [cm]
Straight	S. wheel	3.7
	Digitize	5.0
	Continuous	6.2
Circuit	S. wheel	8.7
	Digitize	10.2
	Continuous	10.6

We think the angle sensor or rotary encoder is one of best devices to be controlled by body parts (as mentioned in Section IV). Until now, it is expensive for disabled people to customize their car. However, they could drive their car if the sensor operation unit was simply connected to its power-steering motor. This operation unit is very flexible and cheap, and their car does not need to be significantly customized.

Figure 9. Operation of load-cell-sensor unit

Figure 10. Operation unit mounted in a car

Figure 11. Right/Left-angle corner used for evaluating turning drivability

VI. CONCLUSION

Several kinds of driving operation units for disabled people have been developed. With one of those units, an angle-sensor operation unit, the driver could drive well on roads with gentle curve. However, this sensor unit was found to be unsuitable for sharp corners. Accordingly, the angle sensor was replaced with a load-cell sensor. It was experimentally shown that the driver can control the angle of the steering wheel by adjusting the intensity of the stepping force they apply to the load cell. By connecting the load-cell operation unit to the power-steering unit on a standard car, it was experimentally shown that the driver can control the rotation speed of the steering wheel by adjusting the intensity of force applied to the unit. A driver could drive the car in a manner close to that achieved with a steering wheel after just a bit more practice than needed for a steering wheel. Moreover, it was found that a device that has seven steps, for example, an angle-sensor operation unit with seven steps, is suitable for enabling physically disabled people to drive a car. Since such a unit is very flexible, cheap, and needs little customization to fit in a car, it would be suitable for physically disabled people.

ACKNOWLEDGMENT

Thanks to Kazuhiro Yoshida, Toshiyuki Sato and Akihiro Itami for help in performing this research. This work was supported by JSPS KAKENHI Grant Number 16K00276.

REFERENCES

[1] Development of Honda's Franz System Car; http://www.honda.co.jp/50years-history/challenge/1982franzsystemcar/index.html, [in Japanese, retrieved: May 2019]

[2] M. Wada and F. Kameda, "A joystick car drive system with seating in a wheelchair," IEEE IECON '09, pp. 2163-2168, November 2009.

[3] S. L. Poczter and L. M. Jankovic, "The Google Car: Driving Toward A Better Future?," Clute Institute , Journal of Business Case Studies, Vol. 10, No. 1, pp. 7-14, 2014.

[4] T. Minato, Y. Murata, A. Suzuki, N. Sato, and Y. Sasaki, "Development of Automobile Driving Interface with Strain Sensor for Disabled People," Information Processing Society of Japan, CDS Transaction, Vol. 1, No. 1, pp. 1-11, 2016 [in Japanese].

[5] PFS, TOYO SOKKI Co. Ltd., http://www.toyo-sokki.co.jp/download/pfs/pfs_catalog.pdf, [in Japanese, retrieved: May 2019].

[6] B. Pfleging, S. Schneegass, and A. Schmidt, "Multimodal Interaction in the Car - Combining Speech and Gestures on the Steering Wheel," ACM, AutomotiveUI '12, pp. 155-162, 2012.

[7] S. Anand, J. Terken, and J. Hogema, "Individual Differences in Preferred Steering Effort for Steer-by-Wire Systems," ACM, AutomotiveUI '11, pp. 55-62, 2011.

[8] B. Sucu and E. Folmer, "The Blind Driver Challenge: Steering using Haptic Cu," ACM, ASSETS'14, pp. 3-10, 2014.

[9] Y. Murata and K. Yoshida, "Automobile Driving Interface Using Gesture Operations for Disabled People," IARIA, International Journal on Advances in Intelligent Systems, vol 6 no 3 & 4, pp.329-341, 2013.

[10] Research Institute for Electromagnetic Materials, http://www.denjiken.or.jp/d-htm/research/group.html, [in Japanese, retrieved: May 2019].

[11] TMI-160, TOYO SOKKI Co. Ltd., http://www.toyo-sokki.co.jp/pdf/TMI-160.pdf, [in Japanese, retrieved: May 2019].

[12] PIUS Kit Car, http://www.pius-kitcar.com/, [in Japanese, retrieved: May 2019].

Proposal of Traffic Database Management System

Yoshitoshi Murata
Faculty of Software and Information Science
Iwate Prefectural University
Takizawa, Japan
e-mail: y-murata@iwate-pu.ac.jp,

Abstract— **Cooperative intelligent transport systems (ITSs) are attracting much research to sustainably improve traffic environments. The European Telecommunications Standards Institute proposed and standardized the Local Dynamic Map (LDM), which is a conceptual data store embedded in an ITS station. It contains topographical, positional, and status information related to ITS stations within a geographic area surrounding the host station. The number of mobile objects such as pedestrians and automobiles is very big. They often travel around wide areas and sometimes travel across countries. These mean that a distributed real-time database system is needed. This paper presents a "Traffic Database Management System" (TDMS) for managing information related to mobile objects along roads in real time in the world. TDMS performs the same role as LDM but is aimed at the entire world, not a local area. Specifically, the TDMS structures and commands are introduced.**

Keywords- ITS; mobile object; traffic; stream database.

I. INTRODUCTION

Managing location and additional information of mobile object, such as pedestrians and vehicles in an integrated way are very useful for avoiding traffic accidents, reducing energy consumption, analyzing international logistics, etc. The number of mobile objects is too big and its coverage area is too wide to manage their location and additional information with a single server.

Cooperative intelligent transport systems (ITSs) are attracting much research interest as a means to sustainably improve traffic environments, such as by providing driving assistance and navigation information. The European Telecommunications Standards Institute (ETSI) proposed and standardized the Local Dynamic Map (LDM) as a cooperative ITS. LDM is a conceptual data store embedded in an ITS station in vehicles, roadside facilities, and ITS centers. It contains topographical, positional, and status information related to ITS stations [1][2]. The coverage area for each LDM server is limited to a surrounding each host station. Therefore, it is difficult for mobile objects to get information from other LDM servers.

The previously proposed concept of a "Cyber Parallel Traffic World" (CPTW) cloud service is aimed to practice driving in destinations before the trip or drive sightseeing virtually at the coming 5G and connected car era [3]. In CPTW, roads, sidewalks, and traffic facilities, such as traffic signals, are basically represented as they are in the real world. Vehicles, pedestrians, and temporary obstacles move synchronously with their real-world counterparts. CPTW is thus a virtual world synchronized with the real world. Its attention area is not local but worldwide. A previous paper introduced the CPTW concept and the technologies used to create 3D polygons of roads and traffic signals from a road database and a rule database. To achieve CPTW, however, a database is also needed for managing information about mobile objects along roads in real time.

This paper presents the "Traffic Database Management System" (TDMS). Its basic role corresponds to that of LDM for type 4 data (highly dynamic). It differs from LDM in that it manages data from all over the world rather than only local data. Although TDMS handles streaming data the same as sensor networks, it is actually a big data system given that the number of traffic-related mobile objects in the world is tremendous. This means that a distributed database system is needed. Many types of distributed database systems have been developed to deal with big data. Google developed the Google File System [4], Bigtable [5], and MapReduce [6] to deal with the large and ever-increasing amount of Web data. In Google's distributed database system, big Web data are assigned to each distributed server. However, vehicles often travel around wide areas and sometimes travel across countries. Therefore, generated stream data should be uploaded real-timely to the server that manages the area where the data was generated. Distributing the stream data for mobile objects to area servers creates several kinds of problems. For example, mobile objects near area boundaries need information stored in neighbor area servers, which is difficult to do. The proposed system would solve such problems.

In this paper, the concept of TDMS, requirements, database structure, and commands are introduced. The nodes in the system are vehicles and pedestrians. The database stores information about their locations, speed, direction, etc., enabling drivers to obtain information about the mobile objects in their vicinity. Obviously, if drivers are continually accessing the database to obtain this information, the access load will be tremendous. Therefore, the data upload and download speeds must be significantly higher than those of existing sensor network systems.

Prior to a discussion of related work in Section III, CPTW and TDMS concepts are reviewed and introduced in

Section II. The system design, including the database schemes and commands, are introduced in Section IV. The key points are summarized and future work is mentioned in Section V.

II. CPTW AND TDMS CONCEPTS

Before explaining TDMS concept, a review of the concept of CPTW in which TDMS resides will be presented. In the CPTW, roads, sidewalks, and traffic facilities, such as traffic signals are basically represented as they are in the real world. Vehicles, pedestrians, and temporary obstacles move synchronously with their real-world counterparts. Virtual vehicles can be driven in the CPTW in the same way that actual vehicles can be driven in the real world. Unlike in the real world, vehicles and pedestrians can communicate with nearby vehicles and pedestrians by pointing not at their ID (address) but at their position. In this virtual world, it would be possible to detect whether a vehicle or pedestrian obeyed the local traffic rules.

The virtual structures are constructed by extracting road data from maps covering the world and by gathering traffic rules for each locality. These data are stored in a roads database and a rules database. The real-time data for mobile objects, such as the locations of vehicles and pedestrians. are stored and managed in TDMS. An application program creates 3D roads using data in the roads database, creates traffic signals and traffic signs from the rules database, and plots mobile objects, as shown in Figure 1. CPTW is not only useful to experience and practice driving in foreign countries but also to drive sightseeing.

The concept of TDMS is very simple: a stream database is used to store and manage data for all traffic-related mobile objects worldwide, as shown in Figure 2. TDMS would be implemented in connected vehicles and pedestrians since most traffic-related mobile objects are expected to be connected to the Internet through 5G mobile networks.

Archived data as well as real-time data would be stored in each TDMS server. There are two key types of terminals to access TDMS:

(1) A navigation terminal (client sensor device) in each traffic-related mobile object for sending and receiving data records containing ID, Type of object, Time, Location, Speed, Direction, etc. to and from traffic-related mobile objects in their vicinity and for displaying the information received in real time. Data records are selectable by attributing profiles.

(2) A PC that can gather the data for a designated mobile object, and data record contents are selectable. This means that many kinds of services using such data would be provided. For example, some service would extract ego-vehicles, and compulsory confiscate driving a vehicle, or alert vehicles to existence of ego-vehicles. Another one would analyze international logistics.

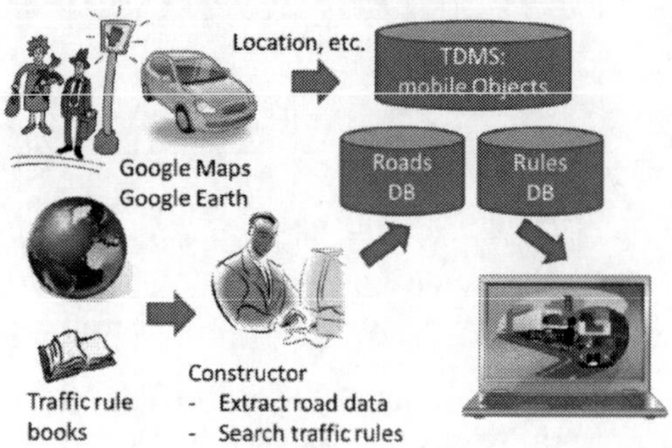

Figure 1. Overview of constructing CPTW

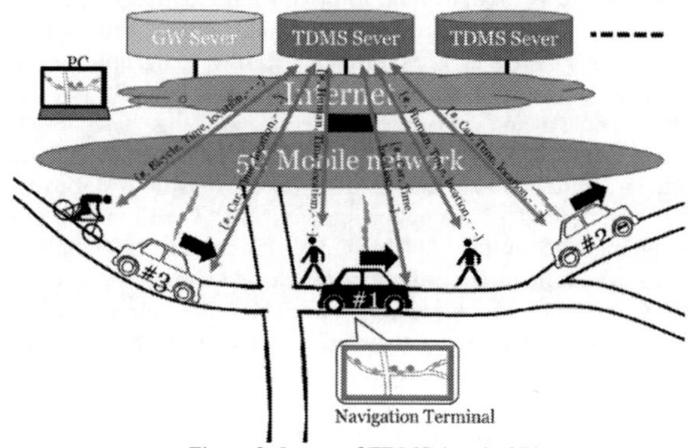

Figure 2. Image of TDMS (not in 3D)

III. RELATED WORK

The worldwide aspect of TDMS makes it unique. The closest related work is that on LDM type of ITS. In an LDM system, data related to traffic are exchanged between ITS stations through vehicle to vehicle (V2V) or vehicle to infrastructure (V2I) communication and stored in a database at each ITS station. An LDM system is a distributed database system without a central server.

Data related to traffic objects can be categorized into four types:

- Type 1—permanent static data, usually provided by a map data supplier,
- Type 2—transient static data obtained during operation, e.g., changed static speed limits,
- Type 3—transient dynamic data, e.g., weather situation, traffic information,
- Type 4—highly dynamic data, e.g., location of a moving vehicle.

The database tables for each type are related using location information. The roads and rules databases in the CPTW correspond to types 1 and 2. TDMS corresponds to type 4.

LDM systems have been implemented in several ways. NAVTEQ-LDM [7], for example, was implemented using SQLite, which is a relational database management system (RDMS) specialized for light-weight machines, such as smartphones and in-vehicle navigation terminals. The Bosch

Tele Atlas PG-LDM implementation is based on PostgreSQL [7], an open-source RDMS specialized for extensibility and standards compliance. In these systems, data, including highly dynamic data, are stored as RDMS records. Another approach was used by Sato et al. who implemented a LDM as a stream database ("SLDM") [10]. Highly dynamic data are input as stream data and joined with static and/or transient dynamic data stored in tables as RDMS records.

Each LDM in an ITS station covers the local vicinity. In contrast, TDMS is aimed at worldwide coverage, so the LDM approach cannot be applied to TDMS.

IV. System Design

In this section, requirements for the TDMS and a system configuration, database structure, and database commands to realize the requirements are introduced.

A. Requirements and system configuration

The number of traffic-related mobile objects, such as pedestrians and vehicles, is very big. These objects are distributed worldwide, and some of them, such as vehicles, move quickly and widely. TDMS requires movement information on each mobile object in real time to prevent traffic accidents.

The system must therefore be able to

(1) process big data distributed worldwide,
(2) gather the real-time locations of all traffic-related mobile objects,
(3) provide the location and type of all mobile objects in the vicinity of each mobile object, and
(4) provide a PC with selected data for mobile objects in accordance with the command selected.

To meet these requirements, TDMS must be a distributed database system comprising a gateway (GW) server, area servers, and navigation terminals, as shown in Figure 3.

Each area server needs four tables:

- a road table containing the identification of each road,
- a road reference table containing the relationships among roads crossing an area boundary line,
- an intersection table containing the identification of each intersection, and
- a mobile object (MO) table containing the information needed to estimate the real-time location of each traffic-related mobile object.

Information on the road on which each mobile object is moving is useful for narrowing down the objects that could potentially lead to an accident. The road estimation function in each area server automatically re-defines a road on the basis of global navigation satellite system location data and registers it in the road database. If it detects several mobile objects traveling along a new route, it automatically defines the route as a road. If it does not detect any mobile objects traveling along a road for a certain period of time, it automatically removes the road from the road database. A road is defined on the basis of the intersections, as described in the next sub-section. An intersection is defined on the basis of mobile object movements.

As described in Sub-section C, the road reference database is needed to manage the relationships among roads in contiguous areas.

Each mobile object needs a subset of the road table containing information for other mobile objects in the vicinity to enable its navigation terminal to recognize the road on which it is traveling and to access data for other mobile objects traveling nearby. This subset is similar to an LDM system. The road table, road reference table, and intersection table are implemented in an RDMS because their data do not need to be accessed quickly. In contrast, the MO table is implemented in RAM as a primary memory FIFO buffer because its data must be accessed quickly.

The database structure, design issues, and control schemes are described in the following sub-section.

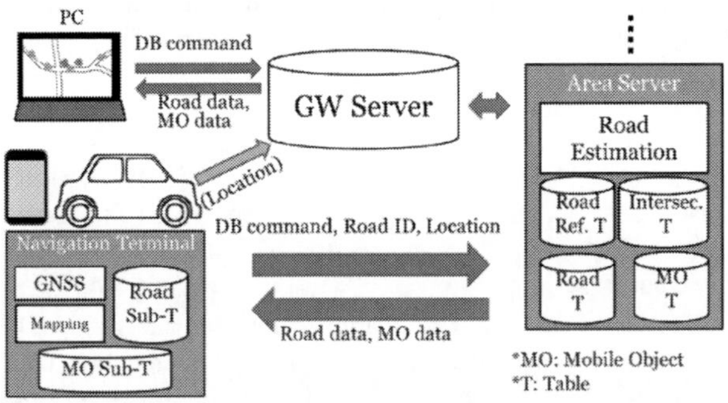

Figure 3. Configuration of TDMS

B. Database structure

Management of related information in the database system is described here.

1) Location format

A scheme in which each area is defined using latitude and longitude is not suitable because it is difficult to represent areas in which a mobile object moves by using two-dimensional values. Instead, the Military Grid Reference System (MGRS) [9] is used. It is the geocoordinate standard used by NATO militaries for locating points on the earth. MGRS code comprises a grid zone, a 100,000-m square identifier, and easting/northing codes, as shown in Figure 4.

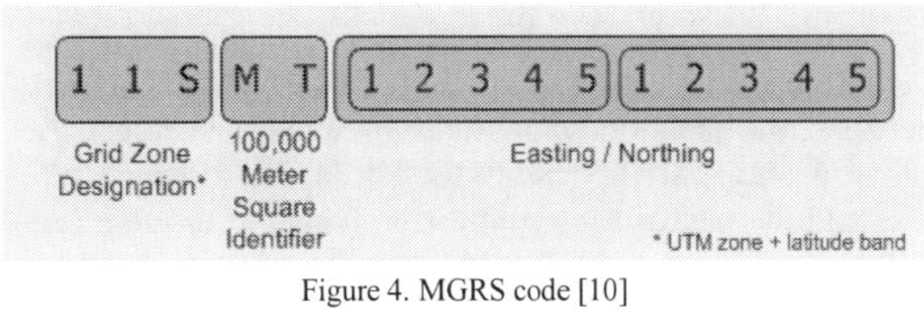

Figure 4. MGRS code [10]

2) Road identification

Traffic accidents occur between mobile objects moving along the same road or on an adjacent sidewalk. Traffic accidents rarely occur between objects moving along different roads even if they are close to each other. Although a vehicle could fall from an overpass and cause an accident below, such events are rare and neglected here. Therefore, it is necessary to estimate only the road on which each mobile object is moving. Since vehicles and pedestrians moving in the same direction sometimes take different directions at an intersection, identifying the road between two intersections is useful for managing traffic-related information. This idea is consistent with that of LDM.

Intersections are sometimes identified with a name or a combination of the names of the intersecting roads. Unfortunately, such identification is not universal. Therefore, in TDMS, intersections are identified by a MGRS point on which vehicles travel to some digitized directions. The road estimation function monitors the movements of mobile objects. When it detects a MGRS point on which vehicles travel to some digitized directions, it registers the point in the intersection table. Since the eight digits of the easting and northing codes specify 10-m square areas, the keys of the intersection table comprise the intersection ID (numeric), the grid zone, the 100,000-m square identifier, the four-digit easting code, and the four-digit northing code. The intersection ID is the primary key in the database.

The keys of the road table comprise the road ID (numeric), the start intersection ID, the end intersection ID, and the address of each FIFO buffer (enabling quick access to the buffers). The road ID is the primary key in the database.

3) Mobile object information

Real-time location information for each object is needed to manage traffic information. Since it would be difficult for each mobile object to continuously send its location information to the TDMS server, it is sent periodically (1 s would be a realistic period). Therefore, to estimate current location of a mobile object, its speed, direction, and time information are needed in addition to the periodic location information.

Furthermore, since some applications need additional information, such as energy consumption, the mobile object (MO) table contains the original grid zone (the first grid zone generated for a mobile object), the type of object (person, automobile, motorcycle, bicycle, etc.), the object ID (automatically assigned to each mobile object in each grid zone, the actual location (main street or side street), the digital location (MGRS code), the speed, the direction, the time, and the payload for application (energy consumption, etc.).

These data sets are stored in a FIFO buffer memory created for each road. The storage of the mobile object database in the practical memory of a server is described in Sub-section D.

Since the location values in these data sets are MGRS codes, the positional relationship between mobile objects can be grasped by comparing the 5th digits of the easting and northing codes in their MGRS code. The positional relationship between mobile objects can also be grasped by comparing the distances from the start intersection to each mobile object. It is difficult to determine in which lane a mobile object is traveling, and the lane information is needed to estimate the risk of an accident. Therefore, MGRS code is used as the location information.

C. *Area division*

Since the data size of TDMS is very big, and traffic-related mobile objects are distributed worldwide, the server should not be configured as a single server system but as a globally distributed server system, like Google Spanner [8]. Each distributed server is assigned a coverage area, as shown in Figure 5. The gateway server and distributed servers for each area could be implemented on the Internet or directly connected to 5G mobile base stations. The latter approach is better suited for exchanging real-time data. That is why the 5G base stations would be distributed worldwide and not all be operated by one operator. TDMS would thus be established on the Internet.

When a mobile object is registered in TDMS for the first time, its navigation terminal should first access the GW server to be assigned an area server. However, if each mobile object accessed the GW server as soon as it started traveling, a heavy load would be placed on the server. Therefore, when a mobile object starts to travel, it compares its current (recognized) location with the (memorized) location where it will finish traveling. If the recognized location is the same as the memorized one, the mobile object sends its information to the memorized area server. If the recognized location differs from the memorized one, it first accesses the GW server.

In the example shown in Figure 5, Japan is divided into nine areas. Area size is decided in accordance with the processing performance of each distributed server. The processing performance required depends on the number of mobile objects traveling in a management area. The average number of mobile objects in each area usually depends on the number of people living in that area. Since the population may change over time, the area division scheme must be flexible.

Dividing the physical areas on the basis of MGRS grid zones would probably be adequate. However, more investigation by simulation in which the practical number of mobile objects is considered is needed to make this decision.

Dividing the physical area into multiple areas causes problems. For example, if the area servers assign road IDs independently, roads crossing a boundary line are registered differently in the two area servers, making it difficult to determine the relationships of the connecting roads. The road reference table solves this problem;

<u>*Road ID assigned by initial area server*</u>*: primary key*
Road ID assigned by next area server

Since the grid zone code for the start intersection in the road ID differs from that for the end intersection for a boundary-crossing road, an area server can extract them. The initial area server sends the grid zone codes for the start and end intersections to the next area server and asks it to find the road for which the grid zone codes for the start and end intersections are the opposite in order to create a road reference database.

For example, as shown in Figure 6, vehicle 1 in area A traveling near the boundary line between areas A and B can obtain data for vehicle 2 but not for vehicle 3 traveling in area B even though vehicle 3 is approaching. A control scheme is needed to solve this problem. The scheme used is for an area server to periodically ask the servers in adjacent areas for data on mobile objects traveling along boundary-crossing roads and on roads that connect to boundary-crossing roads. These mobile objects have a higher probability of colliding with mobile objects in the border area in the near future. Therefore, each area server creates FIFO buffers for such roads and uses to them store data for mobile objects traveling on those roads.

A GW server can access each area server and thereby enable a PC to access an area server through the GW server. A PC can download data using the query commands described in Sub-section D.

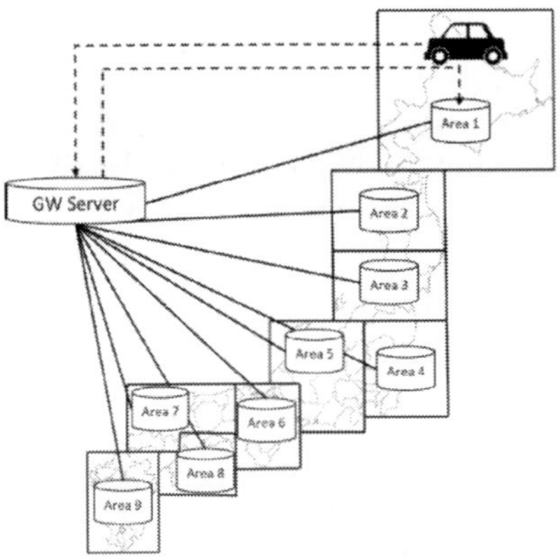

Figure 5. Example coverage of Japan with multiple area servers

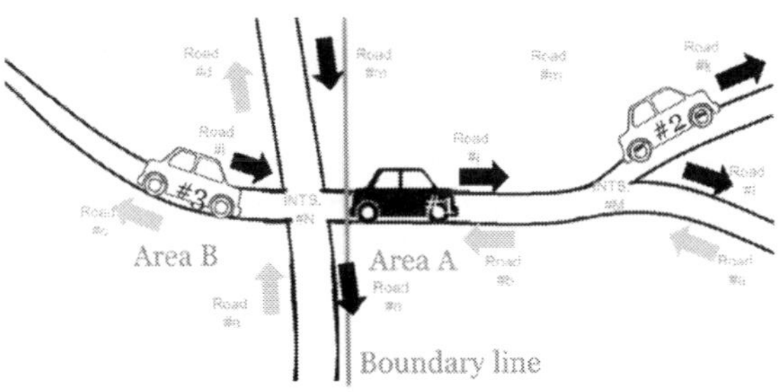

Figure 6. Image of vehicles traveling near area boundary line

D. Storage structure

Since TDMS needs very quick query processing, the memory structure is as shown in Figure 7. A FIFO buffer is created for each road and used to store data sets for each mobile object on that road during the specified period, such as 10 s from the present.

The newest data set is inserted into the first memory area of the buffer, and the latest previous data record is shifted to the second memory area. The remaining previous data records are similarly shifted. Data records flooded from the FIFO buffer are stored in an archive memory in the secondary memory, such as on a hard disk drive. The size of the FIFO buffer is set on the basis of the number of mobile objects during the specified period; it is set to greater than the optimum size during the start-up phase and then gradually reduced to the optimum size.

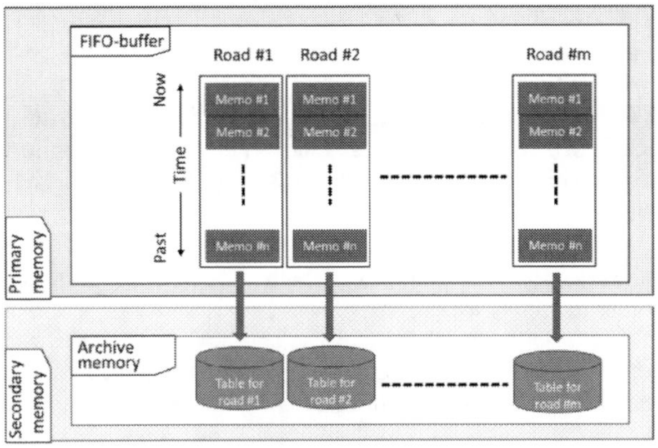

Figure 7. Storage structure in TMDS.

E. Database commands

The following database commands for a navigation terminal are used to insert data records for mobile objects into the FIFO buffer of an area server and for both a navigation terminal and PC to access an area server and extract and download selected data records for mobile objects.

[Create FIFO-buffer]
CREATE FIFO buffer, <road ID>;
used when the road estimation function detects a new road; sent to the management program, which creates a FIFO buffer for the designated road.

[Create other tables]
CREATE TABLE <table> (attribute 1, attribute 2, - - -);
used to instruct road estimation function or management program to creates a road table, road reference table, and intersection table; for road table, the create table command is

CREATE TABLE<road table> (Road ID, Start intersection ID, End intersection ID, Traveling direction, Address of each FIFO buffer).

[Insert data record into a FIFO-buffer]
INSERT FIFO buffer <road ID>
VALUE (v1, v2, v3, v4, v5, v6, v7, v8);
used to insert data sets for mobile objects sent from a navigation terminal into a FIFO buffer.

[Insert data record into other tables]
INSERT INTO <table> (attribute 1, attribute 2, - - -)
VALUES (v1, v2, - - -):
used to insert data into the road table;
INSERT FIFO-buffer <road> (Road ID, Start intersection ID, End intersection ID, Traveling direction, Address of each FIFO buffer).
VALUES (v1, v2, v3, v4, v5).

[Extract data records for mobile objects]
SELECT <MO_1>, <MO_2>, - - -
FROM <area server ID. road ID>,
WHERE <time>, <distance>:

Two types of attributes for mobile objects (MOs) are specified. One is object type (person, automobile, motorcycle, bicycle, etc. "*" means all types of objects. The other is the MO ID.

"time= present" means the newest data. The time can be set by using "T_1<=time<=T_2." where T is in the format "ssmmhhddmmyy."

The distance is calculated from both the object's newest memorized time, location, and direction; "distance < M" means that the distance from its own is less than M. When an area sever is established for each grid zone, the area server ID is the grid zone ID.

For example, for vehicle 1 sending the following query command to an area server, the database management system on the area server identifies every mobile object within 10 m of vehicle 1 in the FIFO buffers for the road on which vehicle 1 is traveling and for the roads connected to that road:

SELECT Automobile
FROM 55R
WHERE time= present and distance < 10;

If vehicles 2 and 3 in Figure 1 are traveling within 10 m of vehicle 1, the data records for 2 and 3 are sent to vehicle 1 from the area server for which the grid zone is 55R.

V. CONCLUSION AND FUTURE WORK

The proposed Traffic Database Management System (TDMS) is designed to store and manage location, direction, speed, and payload data for every traffic-related mobile object worldwide. In this paper, its system design and database schemes to achieve the TDMS were introduced. Database commands that insert data records of mobile objects to the distributed area server, and request it to extract specified data record of mobile objects were introduced. The ability of TDMS to manage information for mobile objects over a wide area will make it useful for not only implementing the previously proposed "Cyber Parallel Traffic World" and for assisting drivers but also for reducing traffic problems, such as congestion, and for reducing energy consumption.

Future work includes evaluating the ability of TDMS to handle the big data necessary to manage information for traffic-related mobile objects worldwide.

ACKNOWLEDGMENTS

This work was supported by a Japanese Society for the Promotion of Science KAKENHI Grant (16K00276).

REFERENCES

[1] ETSI TR 102 863 V1.1.1, Intelligent Transport Systems (ITS); Vehicular Communications; Basic Set of Applications; Local Dynamic Map (LDM); Rationale for and guidance on standardization, 2011.

[2] ETSI EN 302 895 V1.1.1, Intelligent Transport Systems (ITS); Vehicular Communications; Basic Set of Applications; Local Dynamic Map (LDM) , 2014.

[3] Yoshitoshi Murata and Shinya Saito, ""Cyber Parallel Traffic World" Cloud Service in 5G Mobile Networks," Journal of ICT Standardization, River Publishers, Volume 2, No. 2 (November 2014), Special Issue on ITU Kaleidoscope 2014: "Towards 5G", pp. 65-86, 2014.

[4] S. Ghemawat, H. Gobioff, and S. T. Leung, "The Google File System," In Proceeding of SOSP 2013, 2013.

[5] F. Chang, et al., "Bigtable: A Distributed Storage System for Structured Data," In Proceeding of OSDI 2006, 2006.

[6] J. Dean and S. Ghemawat, "MapReduce: Simplified Data Processing on Large Clusters," In Proceeding of SOSP 2007, 2007.

[7] SP3 – SINTECH – Innovative Technologies, "Safespot Integrated Project - IST-4-026963-IP: Deliverable," http://www.safespot-eu.org/documents/SF_D3.5.4_ KeyConceptsAndExploitation_v1.2.pdf. [retrieved: May, 2019]

[8] Kenya Sato, et al., "Stream LDM: Local Dynamic Map (LDM) with Stream Processing Technology," The Science and Engineering Review of Doshisha University, Vol. 53, No. 3, pp. 28-35, 2012.

[9] The UTM Grid - Military Grid Reference System, Natural Resources Canada, https://www.nrcan.gc.ca/earth-sciences/geography/topographic-information/maps/9789, [retrieved: March, 2019]

[10] James C. Corbett et al., "Spanner: Google's Globally Distributed Database," ACM Trans. Comput. Syst. 31, 3, Article 8, pp.1-22, 2013.

Transmission Performance of an Intra-Vehicle Wireless Sensor Network: An Empirical Approach

Ahmed Aladi and Xiao-Hong Peng
School of Engineering & Applied Science
Aston University
Birmingham, UK
email: x-h.peng@aston.ac.uk

Abstract— **Intra-vehicle wireless communications for sensor-Electronic Control Unit (ECU) links can help reduce wiring harness and improve fuel efficiency - a highly demanded feature for new generation cars. To address the challenges from the inherent poor channel quality and harsh working conditions inside and around a vehicle, this work is aimed to investigate wireless channel properties and transmission performance using a testbed involving a car and wireless communication devices, such as ZigBee, Bluetooth Low Energy (BLE) and Wi-Fi (IEEE 802.11x) modules. Results are generated and analyzed from transmissions across passenger, engine and boot compartments under different environmental and operational scenarios including under interference by other RF signals.**

Keywords—intra-vehicle communications; channel property; transmission performance; testbed

I. INTRODUCTION

Modern vehicles have many sensors, such as temperature, proximity, tire pressure and advanced sensors for autonomous control. Conventionally, sensors are connected using wires to Electronic Control Units (ECUs), which are responsible for collecting sensing data and forward it to corresponding output or applications. The communication protocols of the intra-vehicle sensor network are classified according to the transmission speed and sensing function, from class A requiring low data rates (less than 10 Kbps) to class D with relatively high data rates (up to 1 Mbps). The architecture of the intra-vehicle sensor networks is based on the Control Area Network (CAN) protocol [1] which involves wiring to interconnect among sensor nodes, ECUs, execution actuators and the CAN bus.

As the number of the sensors increases, cabling for connecting different parts in a vehicle will be more problematic. Currently, the wiring harness can have about 4000 parts, 40-50 kg of weight, and 1900 wires for 4 km [2], which imposes significant impacts on fuel efficiency, material cost and diagnostic and maintenance issues. Another issue with the wiring is that for some locations inside the vehicle, it is not possible to connect sensors, such as tire pressure sensors with wires.

To address these issues, wireless communication technologies have been examined for applying in this use case e.g.: deploying wireless sensor networks in vehicles to replace wired connections and provide flexibility to the operation of ECUs. However, this deployment is required to meet strict requirements for safety, level of comfort, energy consumption and pollution [3]. It is also required to meet the demands for increasing the number of on-board sensors as the current intra-vehicle network needs to be re-designed for every production cycle [2]. In addition, the design of such a wireless system in a vehicle will have to address the concerns on the channel behaviour and reliability related performance.

Based on a hardware testbed we set up, this paper presents an investigation on the performance of ZigBee and Bluetooth devices in data transmission across different parts of a vehicle, such as passenger, engine and boot compartments. Our investigation will show the efficiency (throughput), reliability (packet loss rate) of various transmission scenarios, without and with interference. We will also show channel properties in terms of path loss and the cumulative distributed function of the received signal strength for the cases examined.

The organization of the paper is as follows. The related work is discussed in Section II. Section III describes the purposes and settings of four different experiments designed for the investigation on an in-vehicle testbed. The test results and their analysis are presented in Section IV, followed by the conclusion in Section V.

II. RELATED WORK

There has been research work reported that utilized the available wireless technologies, such as ZigBee, specified by the IEEE 802.15.4 standard [4], and characterized wireless channels, such as in Ultra-Wide Band (UWB) [5], millimetre wave [6]. ZigBee and Bluetooth Low Energy (BLE) are main candidate technologies for deploying wireless sensor networks inside vehicles due to their low cost and low power consumption. They both use the unlicensed Radio Frequency (RF) 2.4 GHz global band.

A simple but robust model is presented in [5] to characterize the frequency-dependent transfer function of an in-vehicle UWB channel. A large number of transfer functions spanning the UWB band (3–11 GHz) were recorded inside the passenger compartment of a four-seated car and used to model the intra-vehicle channel encountered and understand the behaviour of the channel in this frequency range.

ZigBee uses the same physical and Medium Access Control (MAC) layers defined in the 802.15.4 standard and has the maximum data rate of 250 kbps, while BLE's data rate is up to 300 kbps. BLE has 40 channels separated by 2 MHz: 3 of them are used for advertisements and 37 channels for data transmission. BLE uses the Frequency Hopping Spread Spectrum (FHSS) technique to hop between these channels [7]. ZigBee has 16 channels on the 2.4 GHz band and uses the Direct Sequence Spread Spectrum (DSSS) technique for the air interface. Both technologies face challenges when they are used for in-vehicle applications due to non-light-of-sight, severe signal scattering and interference problems caused by other sources of radio activities [8].

Costa et al. [9] present the channel characterization of a non-line-of-sight in-vehicle wireless communications at 2.4 GHz frequency band, including a signal reflection beam from ground. Helped by 3D EM simulation, the impact of environmental profiles on path loss performance is specified by using static and dynamic on-board measurements.

Similar work also took place on transportation buses where the E-field strength distribution within an urban bus was studied [10]. In this study, multipath propagation and shadowing were considered to enable E-field exposure analysis and determine the function of transceiver's location within the bus.

A research on intra-vehicle channels both 3-11 GHz and the 55-65 GHz frequency bands provided power-delay profiles which exhibit their differences in root mean square value, delay spread, number of resolvable clusters, and variance of the maximal excess delay [11]. The measured and calculated results also indicated a strong level of noise inside the vehicle examined.

Most related work has been focused more on intra-vehicle channel modelling through simulation or tests, while the work reported here was intended to explicitly reveal the transmission performance in terms of efficiency and reliability of existing wireless technologies. This work was carried out in a real-world environment with varied data transmission scenarios, i.e.: with and without interference, to show the potential of the technologies currently available and identify the areas for improvement in future design. In addition, the performance concerned has been examined in three difference compartments across a vehicle, rather than a single compartment reported in other work.

III. EXPERIMENTAL SETUP

Three testing scenarios (in Engine, Boot and Passenger compartments) were used in our experiments based on a small Vauxhall Corsa 2008 car, as shown in Fig. 1. For obtaining the measurements of the Received Signal Strength Indicator (RSSI), the transmitter was placed in the engine compartment and connected to a laptop to transmit data at a rate of 1Hz (1 packet/s) for one minute each run. The receiving node was placed on the car dashboard. Another laptop having a packet sniffer (Dongle CC2531 for ZigBee or CC2504 for BLE) plugged in was placed close to the receiver node to capture the packet sent from the transmitter. The sniffer was used to monitor and log the RSSI of each received packet.

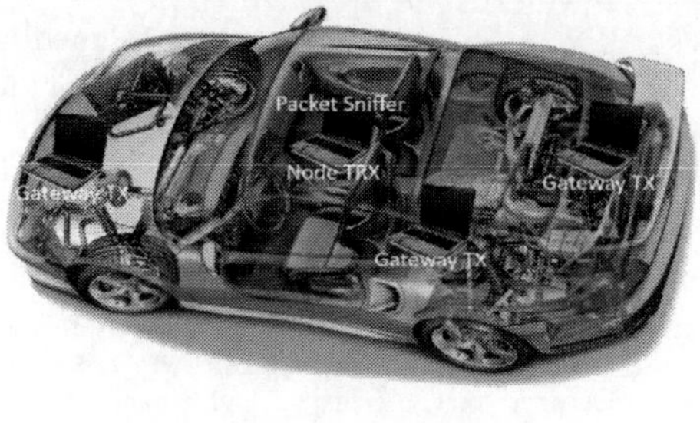

Figure 1. Testbed with testing device positions for three scenarios.

A. *ZigBee Measurements without Interference*

In this experiment, ZigBee transmission was tested for the three scenarios specified above. The receiver was fixed on the dashboard for all scenarios. The ZigBee module used in this research is Digi XBee-PRO S1 802.15.4 with an extended wire antenna. Some of the specifications of this module are shown in Table I.

TABLE I. ZIGBEE PRO S1 SPECIFICATIONS

RF Data Rate	Up to 250 kbps
Receiver Sensitivity	-100 dBm
Frequency Band	2.4 GHz
Interference Immunity	DSSS (Direct Sequence Spread Spectrum)
Transmit Power	0 dBm

Each XBee module was placed on an XBee adapter to provide an easy PC interface for configuring the module using XCTU software. One module was configured as the ZigBee transmitter and the other as the receiver. Both were configured to use Channel 10 and with a transmit power of 0 dBm.

For the throughput and packet loss measurements, the HyperTerminal was used to send out a text file through the XBee module. The file data was transmitted in 4848 packets of size 128 Bytes each. At the same time, the logging software located on the dashboard was capturing the packets from the air.

B. *ZigBee Measurements with Interference*

The impact of the interference from Wi-Fi on ZigBee was observed, given the fact that some Wi-Fi channels have the same frequency as those of ZigBee channels, e.g., Wi-Fi Channel 11 overlaps with five ZigBee channels, 20-24. An ad-hoc connection between two laptops was setup using Wi-Fi and the channel was set to 11 while the ZigBee channel for both the transmitter and packet sniffer was set to 23, to be compatible with the Wi-Fi channel in terms of operating at the same frequency. ZigBee's transmission using HyperTerminal started after setting up the File Transfer Protocol (FTP) between the two laptops.

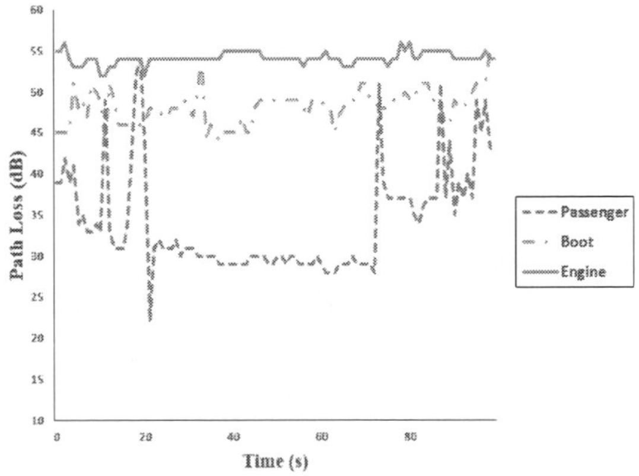

Figure 2. Path loss for ZigBee.

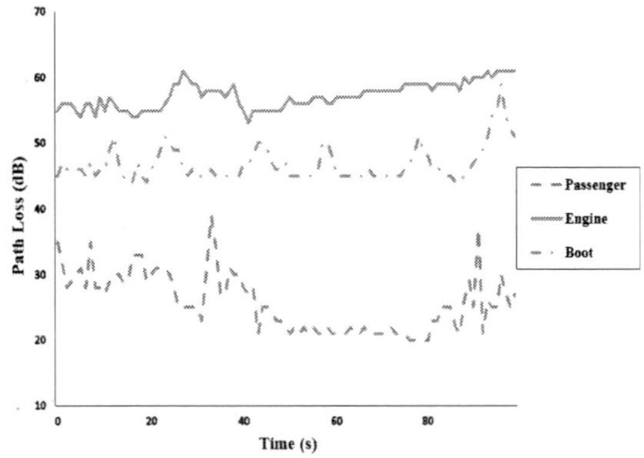

Figure 3. Path loss for BLE.

C. *BLE Measurements without Interference*

BLE has been considered for use in wireless sensor networks inside the vehicle due to its attractive performance in terms of low power, low complexity and low cost [12]. In this experiment, the evaluation of BLE was done by using two BLE nodes, an Android phone that supports BLE, and a BLE packet sniffer attached to a laptop. The phone was used as a sensor node and the CC2540 dongle with a receiver sensitivity of -87 dBm was used as the packet sniffer or receiver. The performances, such as the path loss and packet loss rate were obtained by analyzing the packets captured and applying certain metrics discussed later.

An Android phone was placed inside one of the car compartments, followed by adjusting the transmit power to 0 dBm and the sending rate to 1Hz. The packet sniffer was then used to collect the RSSI value of each packet on 1 second interval up to 100 second in total. For throughput and packet loss measurements, packets were sent at a rate of 7 Hz for a period of one minute. The car was parked in an area without Bluetooth or Wi-Fi signal in order to eliminate any possible interference.

D. *BLE Measurements with Interference*

To examine the effect of coexistence of BLE and Wi-Fi, wireless FTP connection was setup between two laptops. As the BLE frequency is hopping around the three non-overlapping Wi-Fi channels, the Wi-Fi channel was randomly selected from these three channels. The Adaptive Frequency Hopping (AFH) technique specified in BLE was disabled as the transmission was for a single direction only.

E. *Metrics*

We apply the following metrics in this paper for performance evaluation; first, we define some variables:

Packet loss rate (%): $$p = \frac{N-(NR-NF)}{N} \quad (1)$$

Throughput (bps): $$S = \frac{(NR-NF)\times 8B}{T} \quad (2)$$

Path loss (dB): $$L_P = EIRP - P_r \quad (3)$$

where N is the number of transmitted packets, NR the total number of received packets, NF the number of packets that failed Cyclic Redundancy Check, T the total transmission time in second, B the packet size in byte, $EIRP$ the Equivalent Isotropic Radiated Power (transmit power + transmitter antenna gain) in dBm, P_r the received power at the output of the receiver antenna in dBm.

IV. RESULTS AND ANALYSIS

The measurements collected from different experiments will be displayed and discussed in this section.

A. *ZigBee and BLE Transmission without Interference*

The path loss results for both ZigBee and BLE are shown in Fig. 2 and Fig. 3, respectively, which exhibits a very similar channel behavior although measured by different transmission protocols. The Passenger channel varies in a range of 20-30 dB in some instances due to driver movement. The Engine channel has the highest loss among them due to significant signal degradation, caused by multipath fading in such a small enclosure with mixed material, despite the small distance between the engine and the dashboard.

Table II shows the mean and the standard deviation of all the RSSI measurements obtained. The mean of the received signal for the Engine scenario is above the sensitivity threshold of 0.1% of Bit Error Rate (BER) define by both ZigBee and BLE specifications. Both Engine and Boot scenarios seem to have a relatively small variation despite the existence of passengers inside the car.

The Cumulative Distribution Function (CDF) of the measured RSSI is shown in Fig. 4. The specification of BLE [7] mandates a sensitivity better than −70 dBm, but the CDF shows that the probability of the RSSI below that level is almost zero, i.e. even the Engine scenario is also above this standard threshold given the transmit power of 0 dBm.

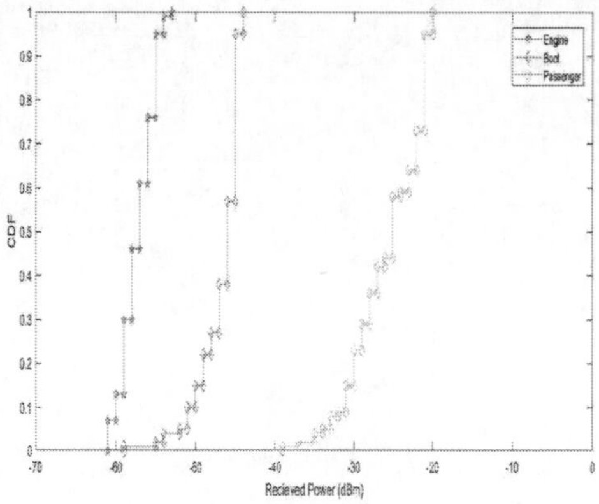

(a) Bluetooth Low Energy

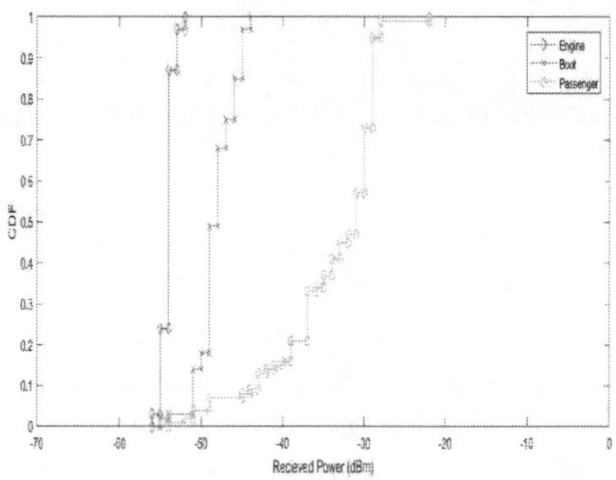

(b) Zigbee

Figure 4. Cumulative distribution function of the recieved power: (a) Bluetooth Low Energy, (b) Zigbee.

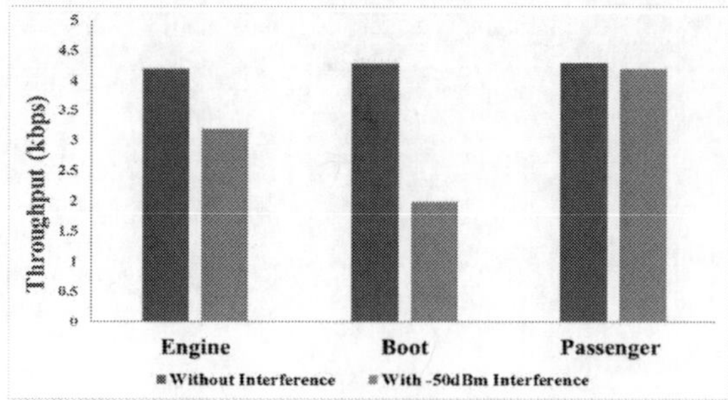

(a) Bluetooth Low Energy

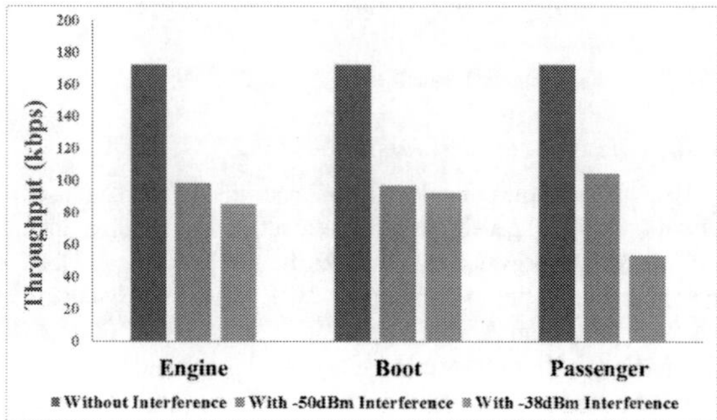

(b) ZigBee

Figure 5. Throughput for all scenarios: (a) Bluetooth Low Engergy, (b) Zigbee.

The throughput performance is shown in Fig. 5, with the maximum throughput being achieved when no interference exists. Fig. 6 also shows that packet loss is not significant for both ZigBee and BLE without interference, although the BLE link has more dropped packets compared to ZigBee.

TABLE II. MEAN AND STANDARD DEVIATION OF RSSI

	ZigBee		BLE	
Scenario	Mean (dBm)	Deviation (dBm)	Mean (dBm)	Deviation (dBm)
Passenger	-34.13	6.44	-25.70	4.49
Boot	-48.17	2.20	-46.83	2.66
Engine	-54.11	0.74	-57.27	1.97

B. Impact of Interference

ZigBee has 16 channels separated by 5 MHz at the 2.4 GHz frequency band, hence any of the non-overlapping Wi-Fi channel with 20 MHz of bandwidth at the same band can overlap with 5 ZigBee channels except Channel 1 in Wi-Fi which only overlaps with 4 ZigBee channels.

In this experiment, Wi-Fi Channel 11 was used to transfer a large file for 2 minutes, and the corresponding ZigBee Channel 23 was used in this case. The ZigBee link suffers from packet losses due to continuous Wi-Fi transmission (Fig. 6). The number of packets dropped increased considerably on a scale of more than 30 %, compared to the zero-interference case. This factor can vary depending on the signal-to-interference ratio (SIR), defined as:

$$SIR\ (dB) = P_r - P_i \tag{4}$$

where P_r is the received signal power of ZigBee/BLE and P_i is the interfering power by Wi-Fi (all in dBm) detected at the receiver.

BLE has 37 data channels with separation of 2 MHz [8]. Only 9 out of 37 BLE channels are free from Wi-Fi interference, which means most of the time the centre frequency during the hopping is overlapping with one of the Wi-Fi channels. This effect can be observed using one of the Wi-Fi analyzer Apps. In this experiment, Adaptive

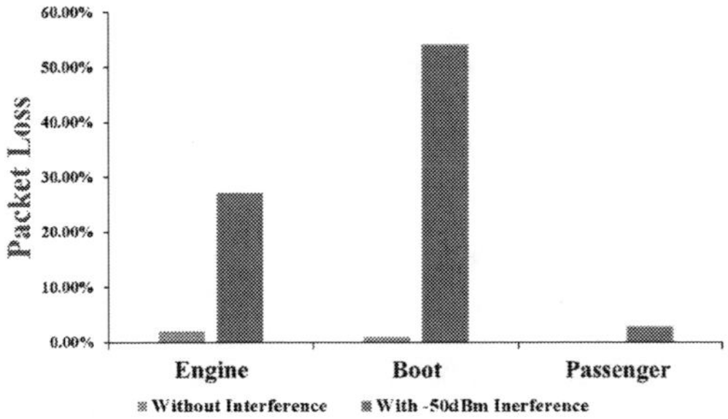

(a) Bluetooth Low Energy

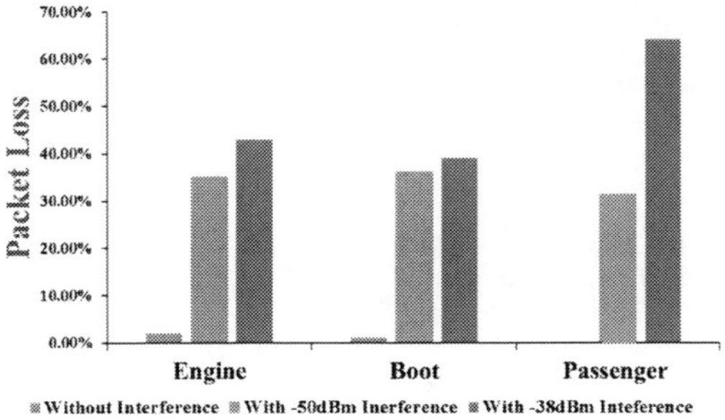

(b) ZigBee

Figure 6. Packet loss for all scenarios: (a) Bluethooth Low Energy, (b) Zigbee.

Frequency Hopping is disabled, i.e., no avoidance of potential interfering channels. As expected in this case, the Boot scenario has a packet dropping rate of 54 %. There is a discrepancy between the packet dropping rates among the three scenarios because *SIR* is high enough for the passenger compartment compared to the other two scenarios. Frequency hopping is operated randomly hence it is difficult to make the same interference period for all the scenarios tested.

C. *Wi-Fi Transmission*

We have also examined the Wi-Fi transmission performance over this testbed. Fig. 7 shows the path loss for each scenario. As expected, the Engine channel suffers a loss around 56-60 dB, more than the other channels. This feature is also reflected in the throughput and the packet loss results, as shown in Fig. 8 and Fig. 9, respectively. Both results are consistent across all scenarios.

There is a significant reduction in throughput for the Engine and Boot scenarios because of the effect of multipath fading on the coherence bandwidth of the channels involved. The Passenger channel can achieve a throughput of up to 52.9 Mbps given the transmitter data rate of 54 Mbps. However, the overall packet loss rate in the Wi-Fi transmission is higher than those in the Zigbee and BLE cases without interference.

The CDF plot in Fig. 10 verifies that the received signal is above the receiver sensitivity (approx. -73 dBm for 54 Mbps) and a reduction in the transmitting power for the Engine scenario could fail this sensitivity threshold.

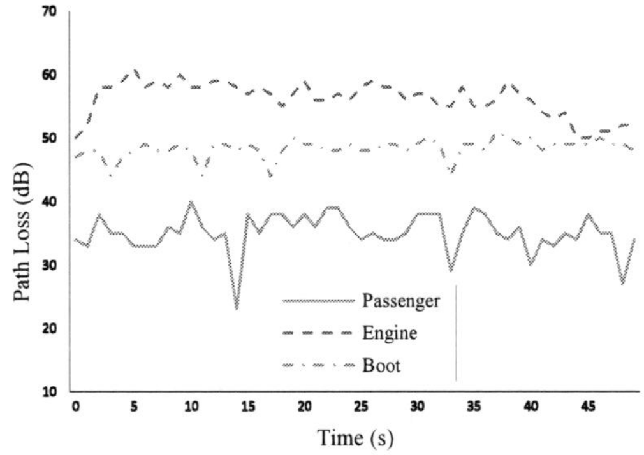

Figure 7. Path loss for Wi-Fi.

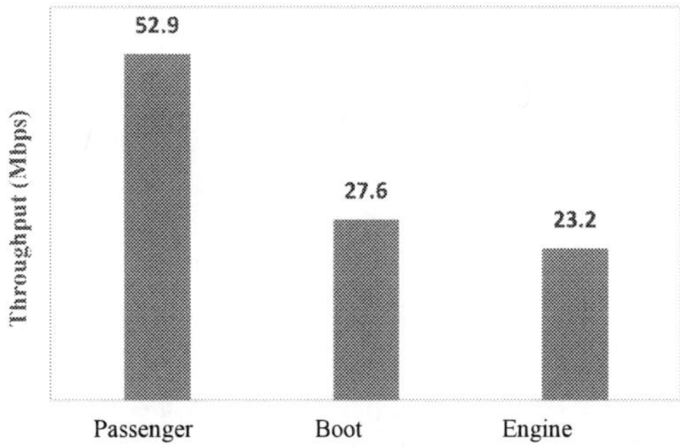

Figure 8. Throughput for Wi-Fi.

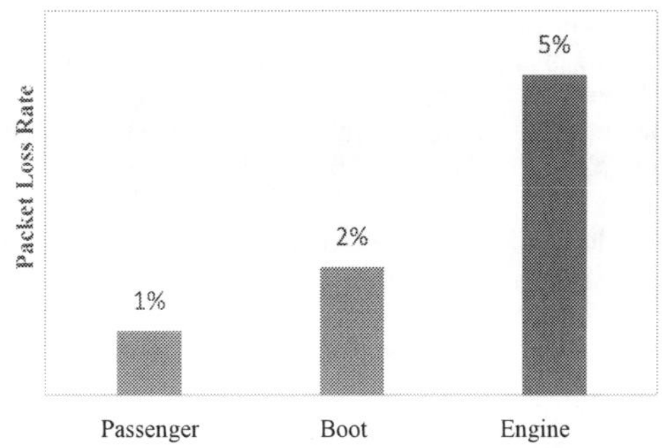

Figure 9. Packet loss rate for Wi-Fi.

The results of these experiments have shown that communications between the transmitter and the receiver in a vehicle can be made reliable provided that some key parameters are adjusted with caution based on the receiver sensitivity specified. The transmit power in the engine compartment needs to be increased to compensate losses, while transmit power reduction can be considered in the passenger compartment to avoid dissipating excessive energy

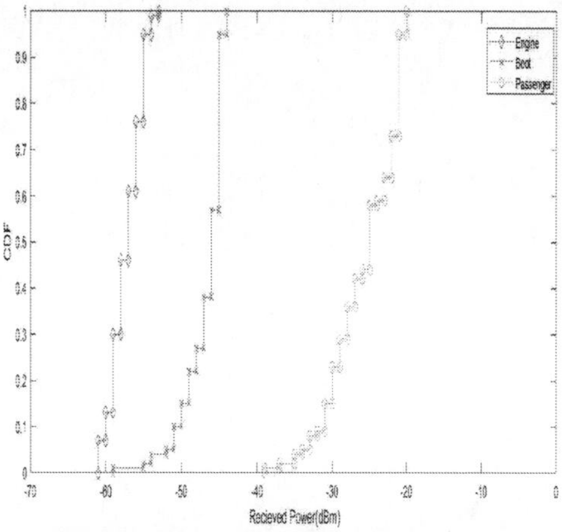

Figure 10. Cumulative distribution function of the recieved power for Wi-Fi.

and causing interference to neighbouring users. In particular, the interference examined in this work exhibits a significant impact on the intra-vehicle transmission performance.

V. CONCLUSION

In this paper, we have investigated wireless channel properties inside a vehicle and the transmission performance of ZigBee, BLE and Wi-Fi as the popular components of intra-vehicle wireless sensor networks. Results have indicated that both ZigBee and BLE can meet the physical layer requirements in terms of the link reliability when they are deployed in any of the compartments in a car. However, the performance can be degraded significantly with co-existence of Wi-Fi transmission, which will lead to serious considerations of the 2.4GHz band in this type of deployment.

To address the problems identified, more robust and adaptive communications protocols and optimization algorithms, such as cooperative communications with virtual MIMO (Multiple Input and Multiple Output) and power control technologies, can be applied to enhance transmission reliability and mitigate the interference encountered.

REFERENCES

[1] Controller Aréa Network Protocol (CAN): "Road Vehicles – Controller Area Network," ISO 11898-1 thru ISO 11898-4

[2] M. Ahmed, C. Saraydar, T. ElBatt, J. Yin, T. Talty, and M. Ames, "Intra-Vehicular Wireless Networks," Proc. IEEE Globecom Workshops, Nov. 2007, pp. 26-30.

[3] L. D'Orazio, F. Visintainer and M. Darin, "Sensor Networks on the Car: State of the Art and Future Challenges," Proc. Design, Automation & Test in Europe, 2011, pp. 1-6.

[4] H.-M. Tsai et al., "ZigBee-based intra-car wireless sensor networks: A case study," IEEE Wireless Communications, vol. 14, no. 6, Dec. 2007 pp. 67–77.

[5] A. Chandra et al., "Frequency-domain in-vehicle UWB channel modeling," IEEE Trans. on Vehicular Technology, June 2016, pp. 3929-3940.

[6] J. Blumenstein et al., "In-Vehicle mm-Wave Channel Model and Measurement," Proc. IEEE 80th Vehicular Technology Conference (VTC2014-Fall), Vancouver, BC, 2014, pp. 1-5.

[7] R. Heydon, Bluetooth Low Energy: The Developer's Handbook. 1st ed. 2012, p.84.

[8] N. Lu, N. Cheng, N. Zhang, X. Shen, and J. Mark, "Connected vehicles: Solutions and challenges," IEEE Internet of Things Journal, 1(4), pp.289-299, 2014.

[9] C. A. M. Costa et al., "Damper-to-damper path loss characterization for intra-vehicular wireless sensor networks," Proc. 47th European Microwave Conference (EuMC), Oct. 2017, pp. 1341–1344.

[10] M. Celaya-Echarri et al., "Spatial characterization of personal RF-EMF exposure in public transportation buses," IEEE Access, vol. 7, pp. 33038-33054, March 2019.

[11] J. Blumenstein et al., "In-vehicle channel measurement, characterization, and spatial consistency comparison of 3-11 GHz and 55-65 GHz frequency bands," IEEE Trans. Veh. Technol., vol. 66, no. 5, pp. 3526–3537, May 2017.

[12] J. R. Lin, T. Talty and O. K. Tonguz, "On the potential of Bluetooth low energy technology for vehicular applications," IEEE Communications Magazine, vol. 53, no. 1, pp. 267-275, January 2015.

Trust in Automation: An On-Road Study of Trust in Advanced Driver Assistance Systems

Liza Dixon
Hochschule Rhein-Waal
University of Applied Sciences
Kamp-Lintfort, Germany
email: lizadixon@gmail.com

William M. Megill
Hochschule Rhein-Waal
University of Applied Sciences
Kleve, Germany
email: william.megill@hochschule-rhein-waal.de

Karsten Nebe
Hochschule Rhein-Waal
University of Applied Sciences
Kamp-Lintfort, Germany
email: karsten.nebe@hochschule-rhein-waal.de

Abstract—**Appropriate user trust is critical in ensuring the acceptance and safe use of Advanced Driver Assistance Systems (ADAS). Despite the prevalence of ADAS on-road today, there is a limited understanding of how trust is affected by a user's first contact with the system on-road. Ten participants without prior experience were introduced to a level 2 system and completed an on-road test drive session. Utilizing a mixed-methods approach including the Trust in Automation (TiA) questionnaire, verbal trust scores, and Facial Emotion Recognition (FER), trust in the system was measured at key milestones. TiA scores increased in a majority of participants, and a significant shift in the factor *Reliability/Competence* ($p<0.05$) was observed post-drive. According to FER scores, participants with a gain in TiA post-drive and those with a loss in TiA post-drive, more frequently displayed the emotions *happy* and *angry*, respectively. Results indicate that trust increases after a user's first experience with ADAS and further that FER may be predictive of user trust in automation.**

Keywords--Advanced Driver Assistance Systems (ADAS); Trust; Human-Machine Interaction; Autonomous Vehicles.

I. Introduction

Foreign, emerging technologies face a proverbial struggle; acceptance. As a safety-critical system that creates a fundamental shift the role of the primary user (driver to passenger), Automated Driving Systems (ADS) present substantial challenges in user acceptance.

The term *Automated Driving System* refers to vehicles that are conditionally, highly or fully automated (level 3, 4 or 5, respectively), whereas, Advanced Driver Assistance System (ADAS) refers specifically to level 1 (assisted) or level 2 (partially automated) systems [1][2]. The pervasiveness of partial automation (level 2) in the form of Advanced Driver Assistance Systems (ADAS) and the dawn of conditional automation (level 3) in production vehicles on-road necessitates a robust understanding of Human-Machine Interaction (HMI) challenges in vehicle automation [1].

From a road safety perspective, the incentive for acceptance is clear—a majority of auto accidents are due to human error, killing 1.35 million people each year and leaving up to 50 million injured or disabled, internationally [3]. In addition to a humanitarian concern, acceptance is also an economic concern for companies heavily invested in the research and development of vehicle autonomy [4]. Failure to support users in their exchange with ADAS on-road today, "will become increasingly costly and catastrophic [5]," as these systems grow in their prevalence and capabilities.

A. Trust in Automation

Studies have been conducted to gather insight regarding user sentiment towards vehicle autonomy, the results of which point to trust as a major factor in the acceptance of the technology [6]–[8]. Trust is defined by Mayer et al. [9] as an attitude, it is not risk-taking, "but rather it is a willingness to take risk." Adjusted for the context of automated systems by Körber, trust is, "the attitude of a user to be willing to be vulnerable to the actions of an automated system based on the expectation that it will perform a particular action important to the user, irrespective of the ability to monitor or to intervene [10]." Trust affects reliance on automation, and in turn reliance aids the user in navigating the complexities of automated systems, especially when the context of use demands adaptive behavior, as is the case with ADAS [5].

1) Closed Courses & Simulator Studies: Experiments to measure trust in vehicle autonomy have been carried out in both closed courses [11][12] and simulators [13][14]. For example, in an experiment with 72 participants, Gold et al. [13] utilized a driving simulator modeling a level 3 system to "investigate how the experience of automated driving will change trust in automation and the attitude of the driver towards automation." A questionnaire was administered before and after a 15-20 minute driving experience. Gaze behavior was also recorded in an effort "to measure a change of trust by a change in [eye] scanning behavior." The results of this study revealed that participants reported a higher level of trust in automation after the driving experience, however gaze behavior could not be established as a "valid measurement."

In an experiment using a Wizard of Oz setup (simulating an automated vehicle), Ekman et al. [12] explored a mixed-methods approach for the assessment of trust during a 15 minute drive on a closed course with 18 participants. The results of the study indicated that "data should not only be collected at the very end of a trial only but be complemented with data collection also during a trial, in particular in relation to events that may influence and contribute to a user's overall experience."

2) Framework & Questionnaire: Researchers have developed frameworks, models and scales for the assessment of user trust in vehicle autonomy. Ekman et al. [11] constructed the Lifecycle of Trust (LCoT) framework, to serve as a tool for HMI design. The LCoT identifies 11 trust-affecting factors throughout the *Pre-Use Phase* (*Implicit/Explicit Information*), *Learning Phase* (all activities from *Entering the Vehicle* to transitions from *Manual to Automated Control*, to *Exiting the Vehicle*) and *Performance Phase* (covering *Continuous Usage, Change of Context & Incidents*). Validation of LCoT factors, specifically through the *Pre-Use* and *Learning Phases* are an area of interest for this study, as it is the most current, comprehensive framework for understanding the development of trust in automation.

Based on empirical research, Jian et al. [15] developed the "Checklist for Trust between People and Automation" a 7-point Likert scale comprised of 12 questions designed for use as a general scale in any area where human-automation interaction occurs. Based on this, the work of Mayer et al. [9], Lee & See [5] and others, Körber [10] developed a refined model of Trust in Automation (TiA) with an accompanying 19-item, 5-point Likert scale questionnaire covering the following factors: *Reliability/Competence, Understanding/Predictability, Intention of the Developers, Familiarity, Propensity to Trust* and *Trust in Automation.* The questionnaire features questions, such as "The system is capable of interpreting situations clearly," (*Reliability/Competence*) and inverse items such as "The system reacts unpredictably," (*Understanding/Predictability*) which correspond to the underlying factors. To the knowledge of the authors, this questionnaire has yet to be applied in a study of trust in partial automation on-road.

3) Facial Emotion Recognition: Facial Emotion Recognition (FER) via the analysis of facial expressions is of growing interest to HMI researchers in the area of automated driving. FER is currently being used as a tool for in-vehicle driver state monitoring (via driver facing cameras) in order to improve user experience [16]. According to Lee & See [5], "Emotional response to technology is not only important for acceptance, it can also make a fundamental contribution to safety and performance."

FER consists of three main events: 1) face and facial component detection, 2) feature extraction, and 3) expression classification [17]. The facial expressions which are associated with the emotions *happy* (*joy*), *anger*, and *surprise* are thought to be the most relevant in the context of automated driving and are used by commercial software companies in their analysis [16]. Studies have confirmed that the emotions *happy* and *angry* are the most influential on how a car is driven [18]. When used as part of a mixed-method approach in post-production, FER enables the observation of connections between driver emotional states, vehicle behaviors and reported trust in automation.

B. Research Questions

While closed courses and simulators provide a stable research environment, they are not fully aligned with the context of use, nor the state of the art. The release of one's personal safety to the system occurs exclusively while engaging with partial automation in an on-road setting. Hence, existing research on is unable to specify how exposure to and experience with a level 2 system might impact user trust. In order to fill this gap in on-road studies, the following research questions were addressed in an experiment:

Q1: How does a driver's first experience with an Advanced Driver Assistance System on-road affect their level of trust in the system?

Q2: What emotions do first time drivers of Advanced Driver Assistance Systems display and is there a relationship between the emotions displayed and their reported levels of trust in the system?

Incorporating a mixed-method approach utilizing verbal trust scores, the Trust in Automation questionnaire [10], Facial Emotion Recognition, and qualitative/observational data, it was expected that the results of the experiment would indicate the following:

H1: Participants will report higher levels of trust in ADAS after their first experiential drive with an ADAS.

H2: FER analysis will reveal a relationship between a participant's TiA score and their emotions displayed during the drive.

Section II of this paper discusses participant demographics, the technical capabilities of the vehicle utilized for test drive sessions, and the experiment procedure. Results are reported in both Section III and Section IV, summarizing qualitative and qualitative findings, respectively. Section V is a discussion of the results, followed by Section VI which outlines the limitations of this study. The paper concludes with Section VII, which offers an outlook on future work.

II. Methods

A total of n=10 participants were introduced to the same level 2 vehicle and completed one individual test drive session. All participants completed their test session within the same two-week period. The driving route was designed as a loop, beginning and ending at the participant's respective campus. The route included driving time on the autobahn (including a construction zone), country roads and in urban settings. Each experiment session lasted 1 hour and 30 minutes, approximately an hour of which was driving time. The same moderator accompanied all of the participants; participants were not explicitly told to activate the ADAS. A pilot test was conducted to refine the experiment structure and equipment, after which it was determined the route did not include enough highway time and was therefore revised.

A. Participants

Of the ten participants selected for this study, there were six females and four males. All participants were members of the university community. The mean age was $M = 31.66$ years ($SD = 9.17$, ranging from 20 to 48 years old). Six of the participants had been driving for over ten years while four had been driving for ten years or less. Educational level was split 50/50 between the participants, half holding a master's degree or above and the other half having received vocational training.

Participants were screened prior to the experiment session to ensure they met specific requirements for the study: holding a valid driver's license, experience with automatic transmission, have no prior experience with the vehicle class (Mercedes-Benz GLC), not own or regularly operate a Mercedes-Benz, have no prior first-hand experience with ADAS, any semi-autonomous or autonomous vehicle systems (including for example: autopilot systems, adaptive cruise control or lane keeping assistants. Excluding: standard cruise control/speed limiters, back up cameras or blind spot assistants).

1) Disclosure: Participants were informed only that they would be taking part in a study of Human-Machine Interaction in ADAS which involved an on-road test drive. The focus of the study being specifically about their trust in the ADAS was intentionally withheld from the participants. It was not disclosed explicitly nor by accident (e.g., titles removed from trust questionnaire, discussion about experiment sessions prohibited) in order to mitigate the *Hawthorne Effect*. This effect refers to the inclination of research participants to adjust their behavior and act in a way that they believe is aligned with the expectation of the moderator [19]. This decision was also made in part due to the high cognitive demands of the experiment [20] (driving an unknown vehicle with unfamiliar technology on public roadways, while under observation) and to obtain unbiased and natural reactions in any participant commentary related to the discussion of trust in the system.

B. Vehicle

The same Mercedes-Benz GLC-250 4Matic was driven by all participants. This vehicle was equipped with the Driving Assistance Package Plus option which includes an Advanced Driver Assistance System (sub-systems relevant to this study are listed in TABLE I.). These features qualify the vehicle as a partially automated, level 2 system [1].

TABLE I. SELECT DRIVER ASSISTANCE PACKAGE PLUS FEATURES.

Feature	Function	Active
Distance-Pilot DISTRONIC with Steering Assist and Stop&Go Pilot	"Autonomous intelligent cruise control system" able to accelerate and decelerate according to traffic conditions. Steering interventions help the driver stay in lane. The system can follow the vehicle ahead even where there are no or unclear lane markings (<130 km/h).	0-200 km/h, driver activated
Hands-Off Warning	A haptic (steering wheel vibration) and graphic warning (in the multi-function display, next to the speedometer), alerts the driver to return their hands to the wheel. If this is not heeded, it is enhanced via an auditory warning tone.	Active with DISTRONIC
Active Lane Keeping-Assist	Detects unintentional lane drift by monitoring road markers. Can tell if the vehicle veers out of lane without signaling, and will vibrate the steering wheel. Brakes individual wheels for correction, keeping the vehicle within the road markers.	60-200 km/h, (conditional)
PRE-SAFE® Brake with Pedestrian Detection	Able to detect pedestrians ahead and will apply the brakes automatically.	Up to 50 km/h
Traffic Sign Assist	Identifies traffic signs and speed limits on the instrument display via camera and GPS data.	Always active.

Source:[21].

The purpose of this study is not to cross-compare various technologies, but rather, to analyze the inherent trust in a particular vehicle's systems, holding this as a constant.

C. Procedure

In order to ensure consistency and objectivity between the experiment sessions, the moderator adhered to a set procedure (see TABLE II.) and script. At the start of the session, the moderator greeted the participant outside of the vehicle in the parking lot. This is when the participant was first exposed to the make and model of the vehicle. The participant was invited to enter the vehicle, where they were then interviewed regarding their initial impressions of the vehicle, Mercedes-Benz, thoughts about ADAS, vehicle autonomy, and their expectations of the system, including their initial feelings of trust in the system. They were then asked to give a verbal rating of trust in the ADAS on a scale from 1 to 5 (1=low, 5=high). Next, they watched an introductory video, featuring original content from Mercedes-Benz, edited to reflect the capabilities of the specific vehicle used for testing. The participant was informed that they were in full control of the vehicle at all times and responsible for obeying all traffic laws and posted signs. Next, the Trust in Automation questionnaire (modified from [10]) was administered to the participant in their native language (German or English, translation from [10]). After, they were encouraged to ask questions, to ensure their understanding of the system's functions and capabilities. They were asked to rehearse how to activate/deactivate the system while the car was parked. Following the introduction, each participant was asked for a second time to give a verbal rating of trust in the ADAS. Trust in Automation, Pre-drive vs Post-drive.

TABLE II. EXPERIMENT PROCEDURE

Pre-Drive		Drive	Post-Drive
Introduction I	*Introduction II*	*Test Drive*	*Closing*
1) Interview 2) VTS #1	1) Intro video 2) TiA #1 3) Interview 4) VTS #2	1) Planned route 2) Think aloud 3) FER	1) TiA #2 2) Interview 3) VTS #3

VTS = Verbal Trust Score, TiA = Körber's Questionnaire for Trust in Automation, FER = Facial Emotion Recognition

As the participants began the test drive with the route pre-programmed into the vehicle's GPS, the cameras were activated. The driver-facing camera was mounted to the windshield to the right of the steering wheel for later FER analysis. The driving scene (roadway ahead), multi-function display, and participant's interaction with the system's interface was were captured by a second camera mounted behind/next to the driver's right shoulder.

During the test drive, the moderator did not give any tasks to the participants other than to follow the route on the GPS.

The moderator played an observatory role, giving instruction only when prompted (e.g., clarifying a system limitation). Participants activated the system only as they felt comfortable, in the appropriate conditions and were encouraged to think aloud [5] while doing so. Participants were asked to state aloud whether they or the car was performing certain actions (steering, braking, acceleration/deceleration) throughout the drive and to share their thoughts on the vehicle's behavior as it occurred. Top speed with ADAS active was recorded for each participant as well as adjustments in posture (positioning of hands, arms and feet on/off pedals). Immediately following the drive, the TiA questionnaire was administered a second time. Participants then completed a post-drive interview and gave a final verbal rating of trust in the ADAS on a scale from 1 to 5 (1=low, 5=high), based on their experience. All interview audio was recorded for later reference.

1) Data Analysis: The responses to the TiA questionnaire were scored following the procedure used by the System Usability Scale [22]. Adjusted for the number of questions, responses were reverse coded, added together and then multiplied by a factor to convert the original scores of 0-68 to a 0-100 value, in order to better identify discrepancies in participant's pre-use and post-use scores (the factor *Familiarity* was removed from analysis, as all participants were selected purposefully to have no prior experience with the technology). A Wilcoxon Signed-Rank Test was used to examine differences in pre-drive and post-drive, reverse coded TiA questionnaire medians. This method was chosen as it is appropriate for the comparison of medians in ordinal data from related groups with a symmetrical distribution [23]. Wilcoxon was performed for all TiA factors together (*Reliability/Competence*, *Understanding/ Predictability*, *Intention of the Developers*, *Propensity to Trust* and *Trust in Automation*) and for each factor's respective set of questions. Friedman's Test (adjusted for ties) was used at to analyze shifts verbal trust scores (pre-introduction, post-introduction and post-drive). Friedman's was selected as the data is ordinal, came from a single group measured at three intervals, and there are no interacting effects between the groups [24], [25]. Statistical analysis and plotting of TiA and verbal trust scores was completed in RStudio [26] using the *stats* [27], *agricolae* [28], and *ggplot2* [29] packages.

2) Facial Emotion Recognition: Driver facing video footage was processed by a convolutional neural network (CNN) with 3 convolutional layers and two fully connected layers (including the output layer). The CNN was trained for the facial emotion recognition of seven emotions [30], however a reduced set of emotions was selected for analysis: *neutral*, *happy*, *surprise*, and *angry* [16][20]. Classification performance using this set of emotions was reported 81% accurate by Mathworks MATLAB 2018b [31], which was used to run the network and output the data in text files. The text files were then compiled, cleaned and analyzed in RStudio.

III. QUALITATIVE RESULTS

Participant commentary from the interviews (pre-intro, post-intro, post-drive) and during the test drives was recorded. This Section includes excerpts from the commentary and participant behaviors recorded. The commentary was transcribed and categorized according to the TiA factors examined by the questionnaire [10], and additionally participant *Driving Style* and *Weather*.

A. Reliability/Competence

Toward the end of the drive, P9 said, "I think the benefits [of ADAS] are clear and undeniable. Every system here is intended to improve safety. I don't think there's any danger posed by the system," and "I was skeptical. Having seen it in action, having felt it under my hands…it is a good thing and I could recommend this kind of system to other people as well. I could talk positively about my experiences on the road. I wouldn't be averse to having this kind of system in my own vehicle."

P4 took back control while in an autobahn construction zone due to discomfort with Steer Assist, stating, "It's keeping us in the lanes but before it was a little bit problematic. It went too far to the right and then it went to the left and then I intervened because I was not sure if it would do it itself."

After the drive, P4 said, "There were a couple of mistakes [with Steer Assist] and it was not too clear to me if it was on or off. I guess that's not the point of the system, that I have to focus more on the [system] than the road. It's not as useful because I have to keep my hands at the wheel anyway." During the course of the test drives the system experienced one malfunction, which occurred during P4's test drive. While driving on a country road, the system drifted the vehicle out of its lane. P4 allowed the vehicle to continue drifting out of lane until half of the vehicle was in the lane of the oncoming traffic before intervening.

B. Understanding/Predictability

After the introduction to the ADAS, P7 stated, "I was curious and skeptical at the beginning but now that I know more about how [the systems] work and what they can do for me, it makes me more confident. I think I may struggle with using them due to a lack of experience. I can trust [assistance] more than a full take-over of my driving."

P7 expressed how unpredictable vehicle behaviors affected feelings of trust, "I felt that I was mostly in control of the system, but not when resuming my settings. I knew I was faster when I last used the feature, so I wanted to use it to accelerate. But sometimes it was much faster than I expected. It was alarmingly fast. I did not trust in the braking after such a strong acceleration."

1) Mental Model: Several participants made comments that revealed changes in their mental model of the system as the drive progressed.

During the test drive, when P5 activated the system, they did not release their foot from the pedal until ~40 minutes in to the drive, which automatically caused the distance control system (Distance Pilot) to become passive, leading P5 to become confused about the system's functionality.

P2 said, "I see the lane is not there, so I will not trust [the system.]" and "It is steering but I want to make sure I keep my hands on the wheel because there is no lane marking right there." P6 expressed, "Maybe the [steer assist] is off because it has to "take some information in to analyze the situation," indicating their perception of the system's functionality. P9 stated, "I thought maybe if I moved to the right a little bit, [the system] would start to see a pattern in the lanes. I was trying to show the car what the lane looks like. I thought maybe if you just adjust the position of the car within these lanes that it could find a pattern within it, that it could orient itself."

P8 mentioned a time they adjusted their mental model and hence behavior, "Most times I felt in control. Except for the two times when the car ahead [moved into the adjacent lane and] turned off the road. I thought the system misinterpreted it a bit. I anticipated it the second time, based on the first time, that it could happen, and my anticipation was right."

C. Intention of the Developers

Participants (P1, P3, P5, P7) said that they felt "safe and comfortable" in the vehicle due to the brand. Several mentioned that they would prefer to drive a vehicle from another brand (P1, P8, P9). After an introduction to the system, P9 said, "I wouldn't say I necessarily feel better about [the system]…I think the developers did their best to create a system that doesn't put people in harm. I have faith in them but there are so many variables on the road that I don't feel comfortable putting my full trust in the system."

a) Implicit Information: P9 referenced stories they have heard in the media: "On the news you'll see, in the United States in particular, where people are very excited about automated driving systems, that someone runs into something because they are not paying attention and then ends up in a fatal accident. That puts me on edge about the whole thing. I am not exploding with excitement [to use the system]."

D. Propensity to Trust

In the initial interview, P1 said, "I am like a dinosaur. I don't have a big trust in the system. I feel safe with things I can see." After an introduction to the system, P1 expressed their feelings towards driving: "If I drive, I am concentrated on it and I like it. I would not like a car to do things for me that I could do myself." P7 stated, "I am curious and also suspicious if it really works. I think that I can do better than the automatic calculations of the system. It is making me curious but also cautious."

E. Trust in Automation

P5 expressed a desire to test this system but at each opportunity they intervened and took back control; in the end, stating, "I can't trust the system so fast. You would have to put a wall out of cushions in front of me before I try that" (referring to the Stop&Go Pilot). After the test drive, P5 stated "It's up to me how much I trust the system. If I would drive with such a system for a long time, I would put more trust in it. But now it's a very big contrast in driving for me."

P6 said, "I am not totally trusting but for me a [verbal trust score of] 4 is very high because I normally do not like these systems. But it is very comfortable to me now."

Throughout the test drive P10 was animated and expressive stating at the end, "It is a bit creepy for me…to trust a car. Normally, you trust a driver. I hate to go by airplane. Because you have to trust someone else. A stranger! But here...you have to trust a car...something with no inside, no feelings! It is only a system, a machine, and you have to trust it. The longer you drive it and the more you get familiar with it…you get a feeling for it and you start to trust the system."

F. Driving Style

During the test drive, P9 reflected on the effect the ADAS might have on driving styles, "I could imagine the system really reducing reckless driving. I don't feel the need to even worry about passing this person. I kind of just feel comfortable letting the car takeover. It kind of takes the pressure off me to take some sort of an action. If driving manual transmission, I would probably be more aggressive right now."

P9 stated they would "feel better" if other cars around them were using ADAS. "There are a lot of really bad drivers. I would know the car is going to adjust to keep them within tolerance limits automatically. I would probably feel better about being in traffic with the person." P10 said, "Maybe I would pay more attention to a car [with ADAS]. Because I know it is new a technology. But you can expect more of what a system would do than a human. A system would work or not work, not be in-between like humans. Maybe these are the cars on the road now that you think, 'Oh that driver drives very correctly.'"

1) Risk-taking: Shortly after expressing their apprehension at the start ("I am very nervous about what we're going to do today"), P3 engaged in repeated attempts to test the system at high speeds while on the autobahn. P3 intentionally drifted over lane markings several times to see if the Active Lane Keeping system would reorient the vehicle properly in the lane.

20 minutes into drive, P8 "provoked" the system, stating, "Now [the system] steered, because I provoked it. I tried to go straighter than I should have into the turn. If you get the angle of the curve wrong, it's nice that someone assists you with it."

When testing the Stop&Go Pilot, P7 said, "The car will stop? Cautiously, I am trying that. I've got my foot over the pedal. The car…the car completely stopped. It is new and weird. Okay, wow. Now it's going again on its own. If I know this is doing the job for me, I feel comfortable in releasing my foot and not keeping it directly above the pedals." P9 stated, "So it's going to stop completely? That's giving me a little bit of apprehension right now. That feeling, do I let it? Cause that's like twenty years of driving experience inching up towards that bumper."

Several of the participants drove through or attempted to drive through roundabouts in urban settings while the system was active. P4 followed a vehicle in front through the roundabout and out of the second exit. P8 also followed a vehicle into the roundabout but intervened as the system accelerated once the vehicle in front exited. P6, P7, and P10 approached the roundabout with ADAS active but intervened.

P4 was the only participant who activated the ADAS near its top speed at 190 km/h.

a) Hands Off: All participants with the exception of P5 removed their hands from the wheel long enough to trigger the hands-off warning graphic and/or auditory warning tone.

Halfway through the drive, while traveling on the autobahn (160 km/h), P1 crossed their arms. P1 also adjusted the position of their headrest at high speed, stating "See, this is something I would do now, because the system is on" as they put both arms behind their head and adjusted the headrest.

P4 stated, "You get a warning to put your hands on, but you don't have to do anything, it's just going on its own anyway." P4 discovered a work-around for disabling the hands-off warning; by briefly nudging the steering wheel slightly from side to side they were able to cease the warnings temporarily. P4 continued to workaround these warnings, through a narrow construction zone.

P3 told the hands-off auditory warning to "shut up" and P4 referred to it as "annoying." P4 received the most hands-off warning notifications of all participants (over 45 notifications).

2) *Risk-aversion:* Participants P1, P2, P5 and P6 did not allow the Stop&Go Pilot to come to a full stop while all other participants did. P2 had the lowest top speed at 130 km/h and possessed the most cautious driving style.

G. *Weather*

One instance of rain occurred which lasted approximately 15 minutes toward the end of P3's test drive. P3 said, "I like the system. I trust the system. But because of the rain I have not such a safe feeling because I don't know this car and I am driving it for the first time. It's not like you just sit here and feel safe, because it's up to all of the things that can happen around you." At the end of the drive, P3 stated, "If there was no rain, I would give the system a [verbal trust score of] 5 because it worked, and it did what it was supposed to do. Because of the rain, I didn't feel so safe, so I will say 4."

IV. Quantitative Results

A. *Verbal Trust Scores*

Participants were asked to give a verbal rating of trust in the vehicle's ADAS on a scale from 1 to 5 (1=low trust, 5=high trust) three times throughout the experiment session: pre-interview, post-introduction, and post-drive.

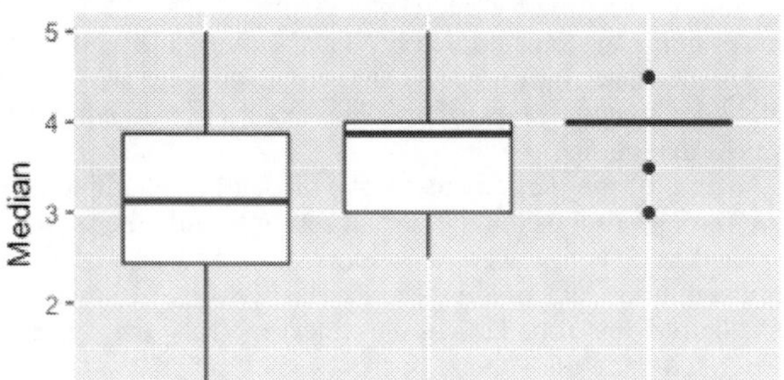

Figure 1. Comparison of median verbal trust scores for all participants at key intervals during the experiment, which were nearly significant at p>0.05.

Verbal trust scores indicated that six of the participants had an increase in trust after receiving an introduction to system (ranging from +0.25 to +1.50 compared to pre-interview scores). Three showed no change and one reported a decrease in trust (-0.50). Post-Drive, seven of the participants reported an increase in trust (ranging from +0.50 to +2.50), while one reported no change and two reported a decrease in trust (-1.0). A non-parametric Friedman test of differences among repeated measures, adjusted for ties was conducted and rendered $\chi2(18)=0.07$, $p>0.05$, which was nearly significant (see Figure 1).

B. *TiA Questionnaire*

The Trust in Automation questionnaire was administered twice during the experiment session: after participants were introduced to the system (pre-drive) and again after the test drive (post-drive). Scored results (see Section II.C.1. *Data Analysis*) from the questionnaire indicated that three participants had a decrease in TiA after the test drive (P1, P5, P4) while all other participants reported an increase of TiA after the test drive (see Table III). All of the participants who reported a decrease in trust after the drive were >30 years of age with ten or more years of driving experience.

TABLE III. Trust in Automation Scores: Pre-drive vs Post-drive.

Participant	**Interval**		
	Pre-Drive	***Post-Drive***	***Change in TiA***
P9	57.408	82.432	+ 25.024
P8	54.464	72.128	+ 17.664
P10	45.632	61.824	+ 16.192
P7	47.104	63.296	+ 16.192
P2	64.8	75.1	+ 10.3
P3	63.3	72.1	+ 8.8
P6	57.408	61.824	+ 4.416
P1	50	47.8	- 2.2
P5	75.1	63.3	- 11.8
P4	54.5	35.3	- 19.2

By participant, pre-drive and post-drive. An increase in TiA occurred in all participants while a decrease in TiA was observed in P1, P4 and P5.

In one instance, a participant's (P1) verbal trust scores did not align with their self-reported TiA responses. P1 verbally reported a gain in trust post-drive, but in the post-drive questionnaire reported a loss of trust.

Each participant's TiA response (based on the 5-point Likert scale) was recoded, and a pre-drive median and post-drive median value was given for each participant. A Wilcoxon Signed-Rank Test performed on all participant's pre-drive and post-drive medians indicated that the post-drive TiA median scores were not significantly higher than pre-drive median TiA scores ($Z=-0.86$, $p>0.05$) (Figure 2, factor: all).

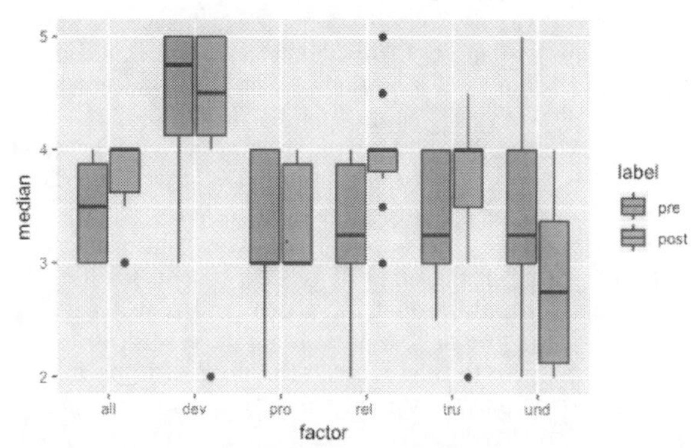

Figure 2. Trust in Automation, pre-drive (pre) & post-drive (post) medians. Shown overall (all) and for each factor (dev=*Intention of the*

Developers, pro=*Propensity to Trust*, rel=*Reliability/Competence*, tru=*Trust in Automation*, und=*Understanding/Predictability*).

The factors *Reliability/Competence*, *Understanding/ Predictability*, *Intention of the Developers*, *Propensity to Trust*, *Trust in Automation* were also considered individually for analysis. The post-drive median score for *Reliability/Competence* was found to be significantly higher than the pre-drive median score (Z=-2.17, $p<0.05$) (see Figure 2, factor: rel). The pre-drive vs. post-drive median scores for the other factors were not found to be statistically significant at $p>0.05$.

C. Facial Emotion Recognition + TiA Score

The driver facing camera footage for each participant's test drive was processed by the convolutional neural network for FER. A value ranging from 0 to 1.0 for each emotional state (where each of the four emotions, *happy*, *angry*, *surprise* and *neutral* share a portion of a 1.0 value) were returned every one tenth of a second for the entirety of the drive. The overall mean values for each emotion were noted separately for each participant. Values were then converted into a percentage, indicating which emotions were most dominant throughout each drive for each participant, respectively. Figure 3 displays the relationship between participant's reported TiA scores and FER scores. The y-axis reflects the change in participants pre-drive vs. post-drive TiA scores, in order from the greatest gain to the greatest loss in TiA. The x-axis presents the FER score as a percentage, indicating the dominant emotion for each participant's drive. *Neutral* was the dominant emotion among all participants, however participants displayed differing frequencies of the emotions *happy*, *angry* and *surprise*. Participants with a gain in TiA post-drive tended to display *happy* whereas participants with a loss in TiA post-drive tended to display *angry* (see Figure 3). Note the distribution of angry among participants and its prevalence in those with a loss in TiA.

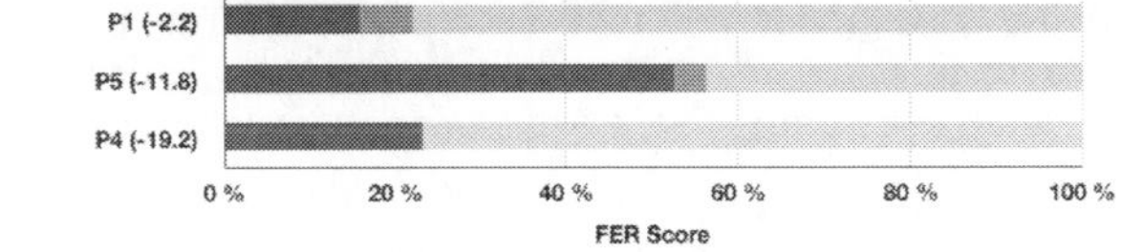

Figure 3. Participant change in TiA Score compared with FER scores (converted to percentage).

According to their TiA scores, participants P9 and P8 reported the greatest gain in trust and (with the exception of *neutral*) displayed *happy* as a dominant emotion. Participants P4 and P5 reported the greatest loss of trust and (with the exception of *neutral*) displayed *angry* as a dominant emotion. A loss of trust was observed in both P4 and P5's verbal trust scores, which decreased at the same interval (both -1.0 post-drive), while P8 and P9 reported an increase in trust at the same intervals (both +1.25 post-introduction and post-drive).

P7 and P10 reported the same gain in TiA post-drive yet displayed different dominant emotions according to FER. A comparison of the verbal trust scores of P7 and P10 do not reveal the same changes in trust (P7 reported verbally no change in trust post-drive, while P10 reported +1.0 post-drive). Additionally, P10 displayed the dominant emotions as those drivers which had the greatest loss in trust (P4 & P5).

Relative to the other drivers in this study, P4 possessed the most aggressive driving style (e.g., speed, triggering of hands off warning graphic/tone). A contrast to P5, who was the most reluctant to give over control to the system and did not take their hands-off the wheel. However, both P4 and P5's TiA scores revealed the greatest loss of in trust in automation post-drive. This loss of trust is also reflected in their verbal trust scores, as they were the only participants to verbally report a decrease in trust post-drive. Further, P4 and P5's both displayed *angry* as a dominate emotion.

V. Discussion

Nesting automation in safety-critical systems requires careful consideration from an HMI perspective. In order to determine what affect a user's first contact with an ADAS has on their level of trust in the system, an on-road experiment with ten participants with no prior experience with ADAS was conducted. The use of the Trust in Automation questionnaire [14], verbal trust scores, Facial Emotion Recognition, and interviews/observational data, enabled a mixed method analysis of each user's experience. It was hypothesized that participants would report higher levels of trust in ADAS after their first experiential drive, and that FER results would reveal a relationship between a participant's TiA score and the emotions displayed (*happy*, *angry*, *surprised*) during the drive.

The scored results of the TiA questionnaire revealed that, trust in automation increased after the test drive in a majority of the participants. A comparison of pre- vs. post-drive median TiA scores however, did not reveal a statistically significant difference in trust in automation ($p>0.05$).

The significant rise in factor *Reliability/Competence* ($p<0.05$, Figure 2) after the drive indicates that based on their experience driving with the ADAS, participants believe that the system performed in a way that reliably assisted them in achieving their goals [5]. Additionally, participants made comments corresponding to the underlying factors of trust in automation, for example, P1 referred to themself as a "dinosaur" regarding their approach to technology, which may be interpreted as an indicator of their *Propensity to Trust*. Reviewing P1's median TiA score for the factor *Propensity to Trust* reveals a low score (pre-drive: 3, post-drive: 2). P9 mentioned their, "faith in the developers" during the experiment session. P9's median TiA score for the factor *Intention of the Developers* was high (pre-drive: 5, post-drive: 5).

Perhaps inherent trust in the established Mercedes brand had an effect on the initial scores, but it is not clear what effect it might have had on the scores post-drive. As noted, a system malfunction occurred during the later part of P4's test drive. One can assume that this was weighed as a factor in their reported feelings of trust in the system. Additionally, it is plausible that the *Hawthorne Effect* [19] may have occurred in the instance where a participant (P1) reported a increase in trust verbally post-drive, but a decrease in trust on the questionnaire post-drive.

Participants with a gain in Trust in Automation post-drive tended to display *happy* more frequently in their FER score while those with a loss in TiA post-drive tended to display *angry* more frequently (Figure 3). This finding is of interest, as self-reported feelings of trust in automation and emotional states appear to follow a similar pattern. This gives validity to the combination of TiA, verbal trust score and FER data, suggesting that this approach may be able to identify a specific persona, who may be less trusting and therefore less accepting of ADAS. However, due to some discrepancies (see Section IV.C.), additional research is needed to determine if a relationship between trust in automation and emotions captured via FER can be replicated.

The driving behaviors of the participants demonstrated a willingness to take risks with the system, for example: using Stop&Go Pilot, hands-off events, and attempting to use the system in complex scenarios such as construction zones, roundabouts, and urban settings. This is aligned with the definition of trust by Mayer et al. [9]. Based on the results of this study however, displaying a willingness to take risks with the system alone is not a reliable indicator of trust as the participant who took the most risks with the system (P4) reported the greatest loss of trust post-drive.

Referring back to the LCoT framework by Ekman et al. [11], the *Learning Phase* events, *Control Transition 1, Automated Mode* and *Control Transition 2* (handover scenarios) do not list *Mental Model* as a trust-affecting factor. This is contradicted by the observations in this study. For example, during the test drive, while thinking aloud, participants stated their beliefs about how the system would behave prior to engaging in *Control Transition 1*. While in *Automated Mode* participants stated their expectations of the system's behavior. When the system behaved in way that was not aligned with their expectations, participants engaged in *Control Transition 2*. Participants then stated why they took back control, and based on their learning from the scenario, adjusted their mental model to adapt their future interaction with the system (see Section B.1). Participants who more easily developed an accurate mental model aligned with the functionality of the system handed control over to the system more easily whereas those whose mental model was not well aligned with the system had a difficult time handing over control to the vehicle. For example, P5 was reluctant to release their foot from the accelerator for an extended period of time, indicating they possessed a poor mental model of the system's functionality. After several *Transitions*, P5 made adjustments to their mental model and their interaction became more fluid. In contrast to P9, who gained an understanding of the system functionality quickly, expressed a desire to work with the system by showing "the car what the lane(s) looks like." P9's accurate model of the system (that the vehicle is tracking the lane markings) allowed for them to place more trust in the system. This suggests that during the *Learning Phase,* both *Automated Mode* and *Control Transition* events are impacted by the trust-affecting factor *Mental Model*.

This study confirms Ekman et al.'s [12] conclusion that a mixed methods approach is required to understand trust in automation. Results also suggest that the finding by Gold et al.'s [13] simulator study ("driving experience increased self-reported trust in automation") does in fact carry over to the on-road context of use. A method for correlating FER data and TiA scores is presented which may be explored in future studies.

VI. Limitations

Studies in simulators and on closed courses allow for significant control over research conditions, whereas studies on public roads leave much open to chance. Each participant was exposed to a variety of different scenarios at varying frequencies and intervals throughout the drive, with unplanned events such as rain or a system malfunction occurring simply by chance. While this study provides insight into the development of trust in ADAS on-road, one should be cautious in generalizing the results of this study, understanding that the it is specific to the design of Mercedes-Benz ADAS.

VII. Conclusion and Future Work

An enhanced understanding of the exchange between the user and the system on-road and the resulting effects on trust, will aid in the design of safer, more efficient automated systems. Further analysis of driver emotional response and/or behavioral cues in correlation with specific driving events and vehicle behaviors (i.e., hands-off, overtaking a vehicle) are of interest due to the relationship between emotion and TiA scores observed in this study. FER accuracy could also be improved by means of data augmentation or by training the network on a more robust dataset. Analysis with an emotion recognition method specifically for driver speech may also be insightful, especially with the movement towards voice user interfaces in-vehicle. The finding that all participants who reported a decrease in trust after the drive came from a similar demographic (>30 years of age, ten or more years of driving experience) warrants further investigation.

The findings presented here support future research of trust in semi-autonomous vehicles and other applications of human-automation interaction. More research is needed to improve the understanding of the development of trust in automation, in order to aid the user in their acceptance of this safety-critical technology. Tackling issues of trust in ADAS today lays the groundwork for the acceptance of higher levels of autonomy in the future, eventually leading to fewer deaths, less injury, disability and a safer more enjoyable on-road experience for all.

Acknowledgments

Thank you to Herbrand Mercedes-Benz and Mr. Sven Ingenpaß for the generous loan of the vehicle for this project. To the administration and staff of Hochschule Rhein-Waal University of Applied Sciences for their support and to the university staff who participated in this study, thank you. Special thanks to Dr. Claudio Abels for assisting with logistics, Dr. André Frank Krause for his assistance on the CNN for FER, Ms. Sabine Lauderbach for her knowledge of statistics, and to Mr. Mario Laugks and Ms. Hale Kadak for their attention to detail.

References

[1] Society of Automotive Engineers (SAE) International, "Taxonomy and Definitions for Terms Related to Driving Automation Systems for On-Road Motor Vehicles: J3016_201806," 2018.

[2] NHTSA, "National Highway Traffic Safety Administration Preliminary Statement of Policy Concerning Automated Vehicles," *Natl. Highw. Traffic Saf. Adm.*, p. 14, 2013.

[3] World Health Organization, "Global status report on road safety 2018," World Health Organization, 2019. [Online]. Available: https://www.who.int/violence_injury_prevention/road_safety_status/2018/en/. [retrieved: May, 2019].

[4] C. Kerry and J. Karsten, "Gauging investment in self-driving cars," *The Brookings Institution*, 2017. [Online]. Available: https://www.brookings.edu/research/gauging-investment-in-self-driving-cars/. [retrieved: May, 2019].

[5] J. Lee and K. See, "Trust in Automation: Designing for Appropriate Reliance," *Hum. Factors J. Hum. Factors Ergon. Soc.*, vol. 46, no. 1, pp. 50–80, 2004.

[6] N. Hutchins and L. Hook, "Technology Acceptance Model for Safety Critical Autonomous Transportation Systems," *2017 IEEE/AIAA 36th Digit. Avion. Syst. Conf.*, pp. 1–5, 2017.

[7] C. Rödel, S. Stadler, A. Meschtscherjakov, and M. Tscheligi, "Towards Autonomous Cars: The Effect of Autonomy Levels on Acceptance and User Experience," *Proc. 6th Int. Conf. Automot. User Interfaces Interact. Veh. Appl.*, pp. 1–8, 2014.

[8] M. Nees, "Acceptance of Self-driving Cars: An Examination of Idealized versus Realistic Portrayals with a Self- driving Car Acceptance Scale," *Proceedings Human Factors Ergonomics Society Annual Meeting*, vol. 60, no. 1, pp. 1449–1453, 2016. doi: 10.1177/1541931213601332.

[9] R. Mayer, J. Davis, and D. Schoorman, "An Integrative Model of Organizational Trust," 1995. The Academy of Management Review, Vol. 20, No. 3 (Jul., 1995), pp. 709-734.

[10] M. Körber, "Theoretical considerations and development of a questionnaire to measure trust in automation," in *20th Triennial Congress of the IEA*, 2018.

[11] F. Ekman, M. Johansson, and J. Sochor, "Creating Appropriate Trust for Autonomous Vehicle Systems: A Framework for Human-Machine Interaction Design," *95th Annu. Meet. Transp. Res. Board*, pp. 1–7, 2017.

[12] F. Ekman and M. Johansson, "Understanding Trust in an AV-context : A Mixed Method Approach," *Proc. 6th Humanist Conf.*, no. June, pp. 13–14, 2018.

[13] C. Gold, M. Körber, C. Hohenberger, D. Lechner, and K. Bengler, "Trust in Automation – Before and After the Experience of Take-over Scenarios in a Highly Automated Vehicle," *Procedia Manuf.*, vol. 3, no. November, pp. 3025–3032, 2015.

[14] M. Körber, E. Baseler, and K. Bengler, "Introduction matters: Manipulating trust in automation and reliance in automated driving," *Appl. Ergon.*, vol. 66, no. January, pp. 18–31, 2018.

[15] J. Jian, A. Bisantz, and C. Drury, "Foundations for an Empirically Determined Scale of Trust in Automated Systems," *Int. J. Cogn. Found. an Empirically Determ. Scale Trust Autom. Syst.*, no. January 2015, pp. 37–41, 2000.

[16] Affectiva Inc., "Affectiva Automotive AI: Metrics in Affectiva Automotive AI," 2018. [Online]. Available: https://www.affectiva.com/product/affectiva-automotive-ai/. [retrieved: May, 2019].

[17] B. Ko, "A Brief Review of Facial Emotion Recognition Based on Visual Information," *Sensors*, vol. 18, no. 2, p. 401, 2018.

[18] T.-K. Tews, M. Oehl, F. W. Siebert, R. Höger, and H. Faasch, "Emotional Human-Machine Interaction: Cues from Facial Expressions," Springer, Berlin, Heidelberg, 2011, pp. 641–650.

[19] L. Bortolotti and M. Mameli, "Decpetion in Psychology: Moral Costs and Benefits of Unsought Self-Knowledge," *Account. Res.*, vol. 13, no. 3, pp. 1–20, 2006.

[20] F. Eyben *et al.*, "Emotion on the Road—Necessity, Acceptance, and Feasibility of Affective Computing in the Car," *Adv. Human-Computer Interact.*, pp. 1–17, Jul. 2010.

[21] Daimler AG, "Active safety: Intelligent Drive: Assistance in all driving situations," *Daimler Global Media Site*, 2019. [Online]. Available: https://media.daimler.com/marsMediaSite/en/instance/ko/Active-safety-Intelligent-Drive-Assistance-in-all-driving-situations.xhtml?oid=10001778. [retrieved: May, 2019].

[22] J. Brooke, "SUS - A quick and dirty usability scale," *Usability Eval. Ind.*, vol. 189, no. 194, pp. 4–7, 1996.

[23] Lund Research Ltd, "Wilcoxon Signed Rank Test in SPSS Statistics," *Laerd Statistics*, 2018. [Online]. Available: https://statistics.laerd.com/spss-tutorials/wilcoxon-signed-rank-test-using-spss-statistics.php. [retrieved: May, 2019].

[24] M. Berenson, D. Levine, and T. Krehbiel, *Basic Business Statistics: Concepts and Applications*, New Jersey. Upper Saddle River: Prentice Hall, 2012.

[25] Statistics How To, "Friedman's Test / Two Way Analysis of Variance by Ranks," *Statistics How To*, 2014. [Online]. Available: http://www.statisticshowto.com/friedmans-test/. [retrieved: May, 2019].

[26] RStudio, "RStudio: Integrated development environment for R." Boston, MA, 2018.

[27] R Core Team, "R: A language and environment for statistical computing." R Foundation for Statistical Computing, Vienna, Austria, 2018.

[28] Felipe de Mendiburu, "agricolae: Statistical Procedures for Agricultural Research." R package version 1.2-8 , 2017.

[29] H. Wickham, "ggplot2: Elegant Graphics for Data Analysis." Springer-Verlag, New York, 2016. ISBN: 978-0-387-98141-3.

[30] I. J. Goodfellow *et al.*, "Challenges in Representation Learning: A report on three machine learning contests."

[31] MathWorks Inc., "MATLAB." 2018.

The Design of a Divide-and-Conquer Security Framework for Autonomous Vehicles

Abdelkader Magdy Shaaban
Center for Digital Safety & Security
Austrian Institute of Technology GmbH
Gieffingasse 4, 12210 Vienna, Austria
Email: abdelkader.shaaban@ait.ac.at

Christoph Schmittner
Center for Digital Safety & Security
Austrian Institute of Technology GmbH
Gieffingasse 4, 12210 Vienna, Austria
Email: christoph.schmittner@ait.ac.at

Arndt Bonitz
Center for Digital Safety & Security
Austrian Institute of Technology GmbH
Gieffingasse 4, 12210 Vienna, Austria
Email: arndt.bonitz@ait.ac.at

***Abstract*—The vehicular security engineering process endeavors to build up a secure vehicle with a high level of security assurance. The well-identified security flaws and conforming security countermeasures help to deliver secure vehicles. This work presents a newly Divide-and-Conquer security framework which can be integrated with the early stages of the vehicular development process to emphasize the security-by-design. The framework proposes to divide the vehicle components into separate layers and sublayers, according to common security parameters. Subsequently, the framework applies a series of security management actions to define potential threats and security vulnerabilities in a vehicle; thereupon, it selects a list of security countermeasures which can mitigate the vehicular's risk. Eventually, the framework performs a security verification and validation to ensure that the vehicle has been developed according to the highest degree of protection level.**

Keywords–Threats; Vulnerabilities; Security Requirements; Risk Assessment; Ontologies; Automotive.

I. INTRODUCTION

The technology of autonomous vehicles is one of the leading innovative research topics in the automotive industry. This technology considers one of the most vivid application examples of the new Internet of Things (IoT) applications. The global automotive IoT market is expected to reach $106.32 billion by 2023 as declared by Netscribes market research [1]. Autonomous vehicles will play an essential role in lessening accident rates and improving traffic efficiency by providing information about traffic conditions, and critical situations. According to the World Bank, traffic congestion can cost developing economies up to 5%, and for developed economies 0.5-3% of their annual Gross Domestic Product (GDP). Therefore, the traffic issues cause to reduce the global economy to $1.4 trillion annually [2].

Fully or highly autonomous vehicles require the cooperation of all road transport actors, road infrastructure, and service providers. These parts influence as a comprehensive infrastructure system that requires new reliable communication approaches to enable communication between vehicles-to-vehicles and vehicles-to-infrastructure. Accordingly, reliable connectivity is the primary requirement for processing various states of the motorized vehicle and accelerating further development [3]. Modern cars can communicate with smartphones via Bluetooth for various purposes, such as hands-free calls, navigation, or multimedia applications. Additionally, new motor cars can connect to the internet to provide additional services such as unlocking and starting the car remotely [4]. These connectivity methods come at a cost; however, that launches a new set of cybersecurity attacks. Other interconnected interfaces, ports, units, or wireless sensors which are directly connected to the internal bus of the vehicle, that can lead to severe attack surfaces [2]. The connected cars and the existence of hackers are now part of life. Therefore, the security must be involved as an integral part of all vehicle development phases to be able to address security vulnerabilities in the early stages of the vehicular development process [2].

Currently, there is no specific risk management framework available for the automotive domain [3]. This contribution presents the first steps into a comprehensive risk management framework for the current and future vehicular industry. The framework proposes to integrate with the vehicular development lifecycle. Divide-and-Conquer inspires the concept of this framework. The Divide-and-Conquer works recursively by breaking down a problem into sub-problems of equivalent specifications until it becomes simple to be solved. The framework follows the same concept of the Divide-and-Conquer by dividing the vehicle into separate layers and sublayers according to common security parameters. Then, the model performs multiple actions on each layer recursively by identifying assets' potential threats, and vulnerabilities. Then, the model evaluates the risks to differentiate between hundreds or thousands of risks that needed to be addressed by the precise security countermeasures. Finally, the framework verifies and validates the selected security countermeasure and suggests additional security countermeasures which can meet the actual security needs.

The paper is structured as follows; the related work on automotive cybersecurity is discussed in section II. Section III includes the main contribution of this work. The section discusses the structured phases of the Divide-and-Conquer security framework. The framework applied to a self-automated vehicle case study as is presented in section IV. The paper concludes with a summary, conclusion, and presents our plans for future work.

II. RELATED WORK

A baseline definition regarding self-driving or partially automated vehicles has been established by SAE International, which has been founded as the Society of Automotive

Engineers. It defines five levels of self-driving technology [5]. By this definition, level zero until two describe varying levels of acclimatization, ranging from warnings and momentary assistance to brake/acceleration and steering support. One example of a more advanced level two vehicle is the Tesla Autopilot [6], which offers both steering support and drivetrain control, but does not yet fall into the SAE level three until five categories. Here, an operator is not considered a "driver" of a vehicle, even when placed in the drivers seat.

A basis for further automation can be extended and more verbose between vehicles, as well as vehicles and roadside infrastructure. However, full connectivity among vehicles and other roadside elements is still under development phases [3]. As described by [3], the connectivity should follow some coordinated model not only based on the vehicle itself but also with the complete infrastructure. The term Cooperative Intelligent Transport Services (C-ITS) summarises these efforts to create a fully integrated transport system. On the forefront of standardization are, as described in [7], the Intelligent Transport Systems (ITS) standards by the European Telecommunications Standards Institute (ETSI). Also worth mentioning are the Cooperative ITS standards from ISO [8]. First attempts to test the feasibility of these standards and C-ITS have been made with the European Cooperative ITS joint development project, which created the first implementation of such a system spreading across the borders of the Netherlands, Germany, and Austria.

The diversity in communication protocols and heterogeneity of components in vehicles that creates new security threats can exploit vulnerabilities to attack vehicle [9]. The work [10] presents several security vulnerabilities, threats, and suggest a variety of security standards for existing and future vehicular systems. However, these points are suitable for particular security conditions in vehicular systems due to the entirely different attacker motivations, attacker skills, and various potential damages [11]. To cope with that, security objectives have to be defined. The first three objectives are Confidentiality (C), Integrity (I) and Availability (A) [12].

III. The Architecture Model of Divide-and-Conquer Security Framework

The lack of existing security framework in the vehicular sector motivates the ISO and the SAE organizations to propose a novel cybersecurity engineering standard for road vehicles [3]. The standard is still undergoing, and the first version is purposed to be published in 2020 [13].

This contribution looks forward to introducing a new security framework for the automotive domain. That work is a part of the Austrian national security research project "Cybersecurity for Traffic Infrastructure and Road Operators" (CySiVuS) [14]. The framework strives to ensure vehicle development life-cycle:

- Identify the potential threats which threatened the vehicle.
- Define security vulnerabilities that can be exploited by potential threats.
- Evaluate the risks of all detected threats and defined vulnerabilities.
- Address the unaccepted risks with suitable security countermeasures.
- Verify and validate the selected security countermeasures to ensure, they meet the actual security protection level.

The security protection level measures of trust that the Industrial Automation Control Systems (IACS) is free from vulnerabilities. ISA/IEC 62443-3-3 specifies security levels that enable a component to mitigate threats for given security protection level [15]:

- SL 1: Prevent the unauthorized disclosure of information via eavesdropping or accidental exposure.
- SL 2: Prevent the unauthorized disclosure of information to an entity actively searching for it using simple means with low resources, general skills, and low motivation.
- SL 3: Prevent the unauthorized disclosure of information to an entity actively searching for it using sophisticated means with moderate resources and moderate motivation.
- SL 4: Prevent the unauthorized disclosure of information to an entity actively searching for it using sophisticated means with extensive resources and high motivation.

Figure 2 depicts the architectural plan of the Divide-and-Conquer security framework. The framework consists of five main phases (i.e., Security Layers, Risk Analysis, Risk Assessment, Risk Treatment, and Security Assurance). These phases are iterative processes and could be started at any separate stages in the process life-cycle, as shown in Figure 1. The following subsections canvass the task of each phase.

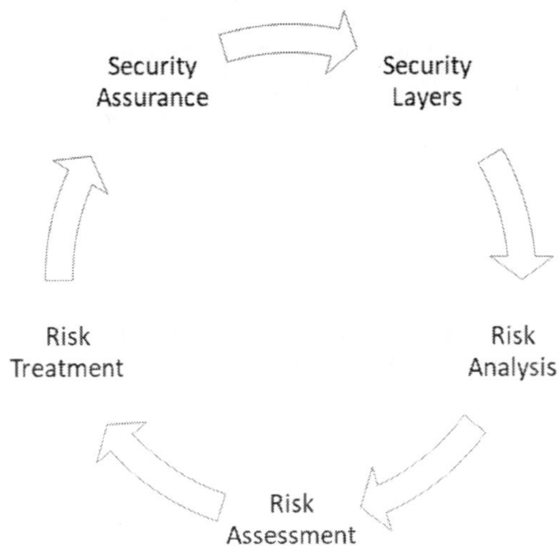

Figure 1. Divide-and-Conquer security framework lifecycle

A. Security Layers

The framework organizes the vehicle into four separate layers. Each layer contains components with common criteria such as type of components, security aspects, security

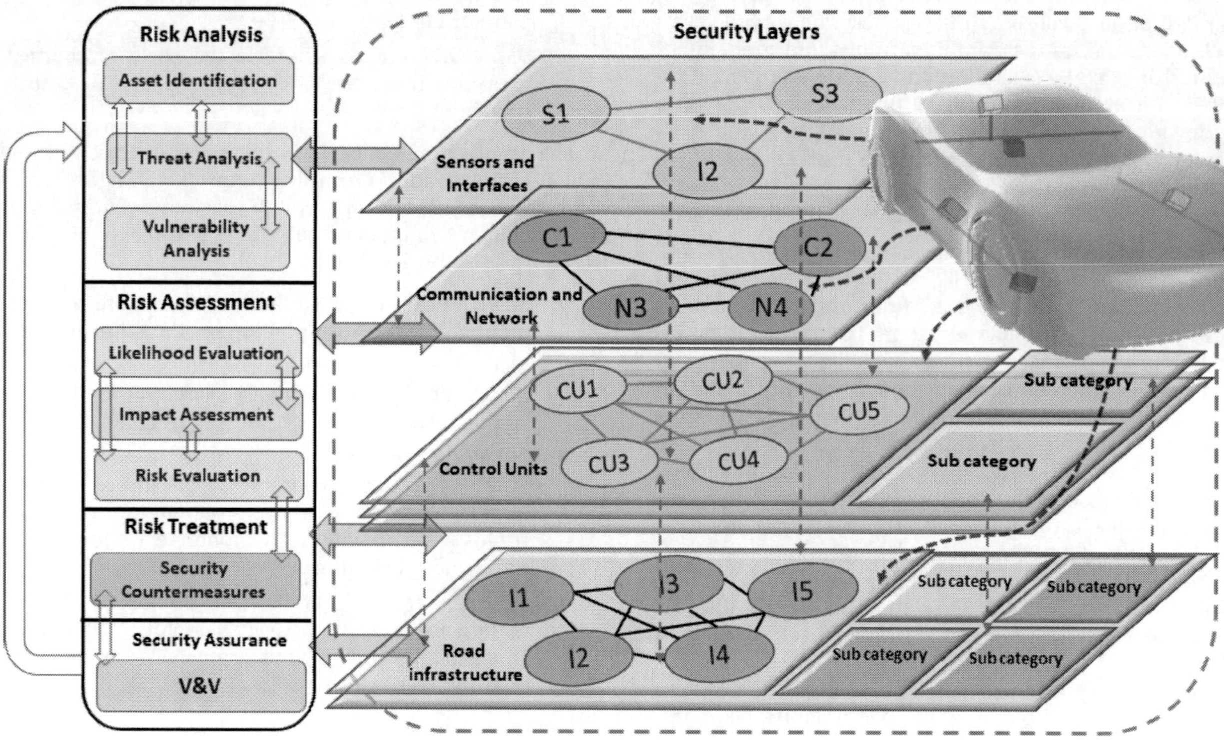

Figure 2. Vehicular security framework architecture model

protection levels, etc. This work proposes to classify the components according to the types of components such as Sensors, Control Units, Actuators, etc. Additionally, some components need more security concerns; consequently, it is proposed to accommodate these components into sublayers according to a specific security protection level.

a) Layer One: Sensors and Interfaces: this layer contains all sensors, ports, and communication interfaces which communicate internal vehicle components together or with the external environment. Some components can be stacked into equivalent sublayers, according to a specific security protection level. For example, internal interfaces which are connected directly with internal units in the vehicle do not need a highly-level of security protection; where external communication interfaces which communicate the vehicle with the external world that require a higher level of security to keep the vehicle secure.

b) Layer Two: Communication and Network: this layer encompasses all communication devices, which allow the internal vehicular units to communicate together within the vehicular boundary. Moreover, the external networking units are organized as highly protected sublayers, which needs more protection concerns.

c) Layer Three: Control Units: this layer has different types of electronic control units (ECUs) in a vehicle. Several ECU units have a list of security parameters (i.e., Tamper Protection, Authentication, Secure Boot, ASIL Rating, etc.) which are used to define the security mechanism for each ECU component. Furthermore, the model uses these parameters to classify the vehicular ECUs according to specific protection level need to be achieved for each unit.

d) Layer Four: Road Infrastructure: as mentioned previously, fully automated vehicles need assistance from the road infrastructure, such as Road Side Units (RSUs), Communication Infrastructure, Traffic Controls, or Intelligent Transportation Systems (ITS) [16]. This layer accommodates all components related to the traffic infrastructure and classifies them into sublayers according to the degree of protection level.

B. Risk Analysis

The risk analysis is an activity that aims to define the relationships between threats and the vulnerabilities which are threatened the vehicle. These relationships establish a set of classes and subclasses to decompose threats scenario into possible attack paths [17]. This activity consists of three main stages as Asset Identification, Threats Analysis, and Vulnerability Analysis.

1) Asset Identification: an asset in a vehicle considered as data, device, component, or either a physical or a logical object. The assets identification process concerns with the following tasks [18]:

- Create an asset record.
- Identify asset information.
- Define the topological structure of interconnected assets.

2) Threat Analysis: in the vehicular domain, the threat analysis is an activity that identifies the potential negative actions that affect the security mechanism in vehicles. The threat analysis process can be divided into the following essential steps:

1) Model the vehicle with all security related assumptions and necessary information.
2) Model potential adversaries with there capabilities, actions, tactics, techniques, and procedures.
3) Apply the threat model to the system model to identify potential threats.
4) Evaluate all identified threats and decide on the risk treatment.
5) Update the system model with the security countermeasures.
6) Repeat step 3 in order to identify missed or new threats.

In the course of our research, we developed the Threat Management Tool (ThreatGet) [19]. ThreatGet identifies and understands potential threats in the automotive domain. It helps to:

- Identify threats.
- Detect security vulnerabilities.
- Evaluate the risks of the identified security issues.

ThreatGet has a threat catalog contains the most common potential threats in the vehicular domain. The threat catalog is managed by ThreatGet to ensure a wide range of potential threats is considered. The following source documents were used to develop the threat catalog:

- Threat Modelling for Automotive Security Analysis [20].
- Connected Cars - Threats, Vulnerabilities and Their Impact [12].
- Threat Landscape and Good Practice Guide for Internet Infrastructure [21].
- A survey of Remote Automotive Attack Surfaces [22].

The tool classifies the potential threats into six main groups according to the STRIDE model (i.e., Spoofing, Tampering, Repudiation, Information Disclosure, Denial of Service (DoS), and Elevation of Privilege) [23].

3) Vulnerability Analysis: the vehicle vulnerability analysis process concerns with exploring, defining, identifying, and prioritizing vulnerabilities or security weaknesses in vehicles. The researcher in the Austrian Institute of technology developed a vulnerability analysis tool is called Failure Modes, Vulnerability, and Effect Analysis (FMVEA) [24]. FMVEA is based on the Failure Mode and Effect Analysis (FMEA) [25] and extends the standard approach with security-related threat modes [26].

C. Risk Assessment

Risk assessment is an activity for evaluating the degree of risks. This activity is based on the parameters of impact and likelihood, which are used to evaluate the specific risk level. This phase takes place after it collects all details about vehicular components, threats, and vulnerabilities which are identified and detected by the previous phase (Risk Analysis). The next subsections discuss the likelihood, impact, and risk evaluation briefly.

1) Likelihood Evaluation: the evaluation of likelihood considers the significant factor for the risk evaluation process. The likelihood assessment is explained in details in [27]. However, this work proposes four different aspects of the likelihood [3]:

- Assumed attacker capabilities.
- Ease of gaining information about the vehicle.
- Reachability and accessibility of vehicular data.
- Tailor-made equipment to attack vehicles.

Table I shows the parameters of the likelihood evaluation. The attacker capabilities are classified according to the skills of the hacker. The data reachability and availability are the main parameters of this evaluation process, which define how easy it is for an attacker to get data. The attacker could use tailor-made devices to attack vehicles; the last row in this Table shows the likelihood values of these outcomes to occur. These parameters are explained in [3].

2) Impact Assessment: the impact assessment process is an activity that aims to evaluate risk when potential threats and security vulnerabilities are defined. In the automotive domain, it is important to ensure that different types of impacts do not damage the vehicle or cause other accident scenarios:

- Causes immediate damage to the environment or human lives (safety).
- Causes the loss of control over personal information (privacy).
- Causes financial damage (finance).
- Negatively impacts the operation and traffic flow (operation).

Table II discusses the impact levels of these four acceident scenarios. The parameter values of these impact levels are discussed in [3].

- Firstly, the direct consequences (i.e., the operational impact would also impact emergency services and could, cause damage to human lives).
- Secondly, assesses the impact evaluation on users and society higher than the impact on the manufacturer. That means the rates of the safety impacts and the financial impacts for users or society are higher prioritized than for organizations. That is because the community trusts the transportation system.

3) Risk Evaluation: this phase uses the estimated parameters of likelihood and impact, as described previously. Then, it performs a risk assessment methodology to calculate the exact risk level. This work applies the well-known risk assessment formula as described in (1).

$$Risk = Threat * Vulnerability * Consequence \quad (1)$$

where:

$$Threat * Vulnerability = \text{Likelihood}$$
$$Consequence = \text{Impact}$$

The formula evaluates the risk level of each detected threats based on the parameters of Table I (likelihood)

TABLE I. PROPOSED LIKELIHOOD PARAMETERS

Parameters	Values			
Capabilities	Amateur (4)	Mechanic, Repair shop, etc. (3)	Hacker, Automotive expert, etc. (2)	Expert etc. (1)
Availability	Public (4)	Information for Maintenance Availability (3)	information for maintenance availability (2)	information for ECUs' company availability (1)
Reachability	Untrusted Network (4)	Private Network (3)	Part time Accessible (2)	Physical Access (1)
Financial	Standard Devices (4)	Specialize Devices (3)	Tailor-Made Device (2)	Multiple Tailor-made devices (1)

TABLE II. IMPACT LEVELS [3]

Impact Levels	User/Society	Manufacturer
Safety	1	-
Operational	3	4
Privacy	2	3
Financial	3	4

and Table II (impact). The results are plotted on a risk scale, is called "Risk Curve," as depicted in Figure 3. This work expects that the Tolerable Value (TV) is equal to two. That means all values (risk evaluation results) above this threshold (TV), need to be addressed by suitable security countermeasure(s) to mitigate risk.

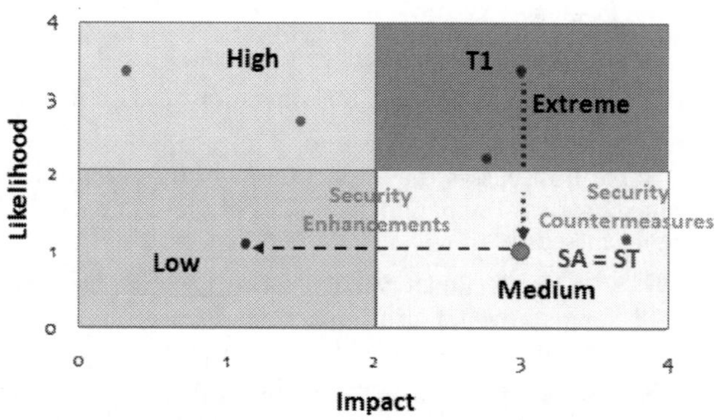

Figure 3. Risk mitigation curve

For example, the point T1 on the Risk Curve is considered as a threat, and its severity is Extreme. Therefore, security countermeasure(s) apply to mitigate that risk into an acceptable security level and to reach the Security Target (ST). Once the security countermeasures are applied, T1 moves into a new level of risk; this state is called Security Achieved (SA), which implies the current state of the security level. The process repeated until the SA=ST otherwise, other security countermeasures, have to be applied to decrease the risk further to a satisfactory security level. In this case, these security countermeasures are called Security Enhancement (SE)

D. Risk Treatment

Selecting the security countermeasures process is considered a significant challenge in the system development process. This phase plays an important role to address potential threats and security vulnerabilities to keep the risk always low. In the course of our research; we developed a Model-based Security Requirement Management Tool (MORETO) for managing a vast number of different security requirements. MORETO reuses different features driven by concepts and knowledge of system modeling.

The tool plays a vital role in this work for generating a list of security countermeasure for a large number of different components. MORETO receives the data generated from the previous phases, such as risk evaluation of the detected threats and vulnerabilities. Then, it generates security countermeasures according to the specific security target which needs to be achieved. MORETO is an Enterprise Architect (EA) plugin for managing the IEC 62443 security standard [28].

E. Security Assurance

In the vehicular domain, the Security Assurance (SA) is a method of guarantee vehicles operate at a high level of security protection. For that purpose, the authors introduced a newly ontology security testing algorithm (OnSecta). OnSecta generates an ontological representation of all detected threats, vulnerabilities, and selected security countermeasures. The algorithm performs a series of inference rules to verify that the selected security countermeasures are handled the detected security flaws. Then, OnSecta validates these countermeasures to a specific security protection level. The algorithm manages numerous security countermeasures which are stored in an ontological structure. OnSecta uses the stored data to select additional security countermeasures when the selected ones have not met the needed security level.

IV. CASE STUDY: SELF-AUTOMATED VEHICLE

Automotive driving is an essential aspect of future transportation. Develop a fully automated vehicle, of the mobility system for people and goods, is demanded. According to the growth of the vehicular industry, new security issues arise. The cybersecurity is considered an integral part of the vehicular development process. A secure vehicle can be designed and developed if security issues are well-identified, and security countermeasures are correctly defined. This case study shows, how our contribution can be conducted in the early stages of the vehicular development lifecycle to ensure security-by-design.

Figure 4 depicts a self-driving scenario. It shows two vehicles A and B communicate together throughout a V2X gateway. The vehicles receive traffic updates from multiple road units over the V2X gateway. The vehicles A and B contain multiple sensors, actuators, electronic control units, internal gateways, various communication protocols, and various communication interfaces. The red links represent the highest risks points in this example, which communicate vehicles with external environments.

All these components are divided into four layers (i.e., Sensors and Interfaces, Communication and Network, Control Units, and Road Infrastructure) according to the Divide-and-Conquer security framework as described in Section III-A. The framework defines the security issues in this example by applying risk analysis methodology as described in Section (III-B). It is essential to evaluate the detected security issues to determine the exact risk, as explained in Section (III-C). Furthermore, it determines the security countermeasure to address the detected security flaws, as discussed in Section (III-D). Finally, the security framework verifies and validates the reached security level to ensure that the SA = ST as considered in Section (III-E). The following subsections explain how the framework applies a series of security methodologies to achieve an acceptable level of security protection and to deliver a secure vehicle.

A. Risk Analysis

This phase applies ThreatGet on that example, and without changs in the security parameters of components, the tool detects over 300 potential threats. Currently, modern vehicles have an average of 60 to 100 sensors, and communication interfaces on board [29]; plus, around 150 Automotive ECUs [30]. Therefore, the number of potential threats in fully automated vehicles is expected to be thousands.

ThreatGet classifies the detected threats according to the STRIDE model. Table III reviews the classes and the numbers of the identified potential threats.

TABLE III. THE CLASS AND NUMBERS OF THE DETECTED POTENTIAL THREATS ACCORDING TO SREIDE MODEL

Threat Types	Numbers
Denial of Service	44
Elevation of Privilege	43
Information Disclosure	88
Repudiation	21
Spoofing	80
Tampering	50

Afterward, the FMVEA tool applied to this example to perform vulnerabilities analysis. The FMVEA defines the vehicles and the road infrastructure as environments. The environment is a container that has components as depicts in Figure 5.

The FMVEA defines security vulnerabilities based on a set of rules to define the structural behavior of components in a given model. In this example, these rules need to be defined first. In this example, the following rules are applied to detect security vulnerabilities:

a) Secure Remote Access Point:
Rule: RemoteAccessPoint.attributes(Authentication = false).hasAncestor(vehicle)
Description: If the remote access point of a vehicle is not secured by Authentication this Access Point could be exploited as weakness in a vehicle.

b) External Gateway Update Frequency and Security:
Rule: Connection.from(InfrastructureGateway) .to(VtoXGateway).attributes(UpdateFrequency>10s, Encryption=false)
Description: External Gateway must communicate in a safe manner over an encrypted connection.

Figure 6 illustrates the detected security vulnerabilities by FMVEA. The red color represents the venerable components in this example,

B. Risk Assessment

This phase evaluates risks based on parameter values of the likelihood and impact level, as explained in Table I and Table II respectively. The assessed risk is classified as one of the primary four risk levels (i.e., extreme, high, medium, or low). The evaluation process focuses only on the highly valued components from the attacker viewpoint, which need a high level of security protection. Figure 7 shows the results of the risks evaluation process of that example.

C. Risk Treatmeant

The MORETO tool plays a vital role in this work to cover the detected security gaps with suitable security countermeasures to mitigate the unacceptable risks. MORETO automatically selects security countermeasures according to the detected threats and vulnerabilities for each affected unit separately, as shown in Figure 8. The figure displays the elected security countermeasures of the V2X Gateway based on the IEC 62443-4-2 security series [31].

D. Security Assurance

The last step is to validate and verify the selected security countermeasures, which are selected by MORETO to coved security flaws in this example. OnSecta is applied to verify and validate the security protection level. It defines the ontological representation of threats, vulnerabilities, and countermeasures, as shown in Figure 9. Then, it applies various reasoning rules to validate and verify the selected security countermeasures and suggests further ones to meet the actual security level.

V. Summary, Conclusion, & Future Work

The paper has introduced a novel comprehensive security framework for autonomous vehicles. The framework aims to be a part of the early stages of the vehicular developments phases to detect security flaws and address these issues with proper security countermeasures. It divides the vehicle into four layers according to the types of components. Each layer contains other sublayers which accommodate components that need a high level of protection. Afterward, the model uses the ThreatGet and FMVEA to define the potential threats and the vulnerabilities in a vehicle. Then, the framework calculates the likelihood and determines the impact levels of the identified security issues. The risk treatment phase selects security countermeasures to mitigate the overall risk. Finally, the OnSecta algorithm verifies the security countermeasures to ensure that all detected threats and vulnerabilities have been handled; additionally, validates the security countermeasures

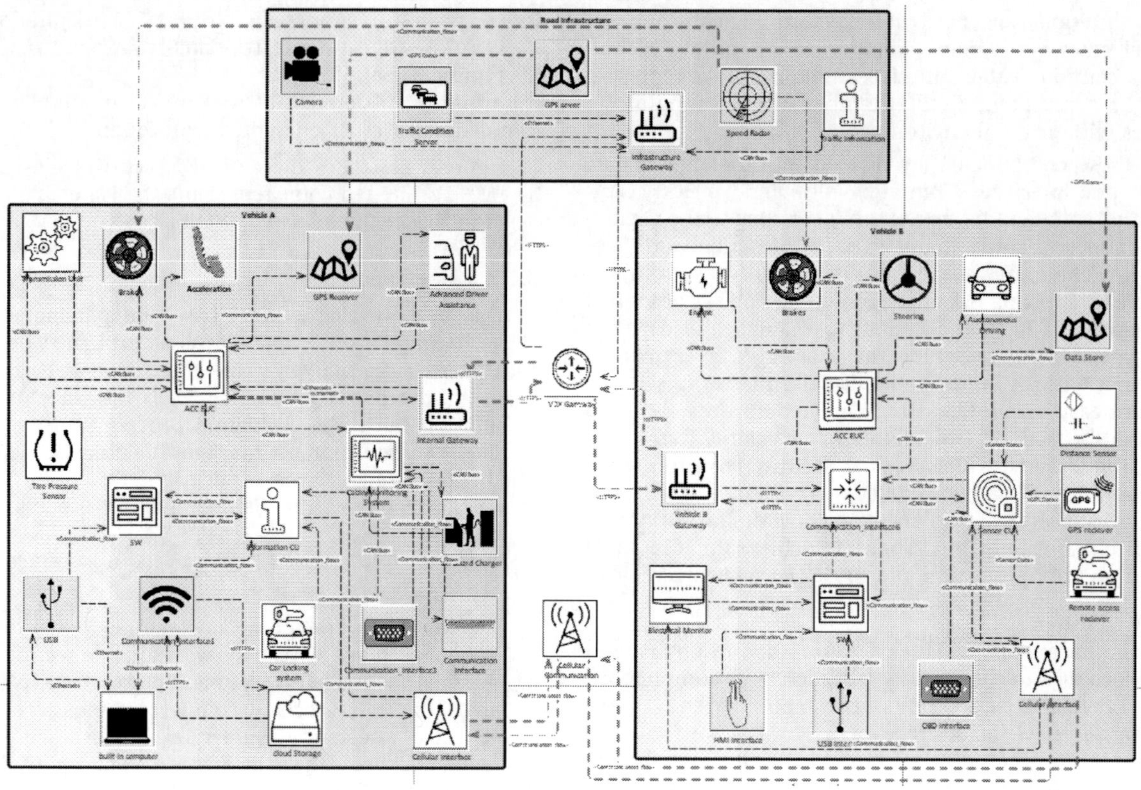

Figure 4. Dataflow between vehicles A, B, and infrastructure units

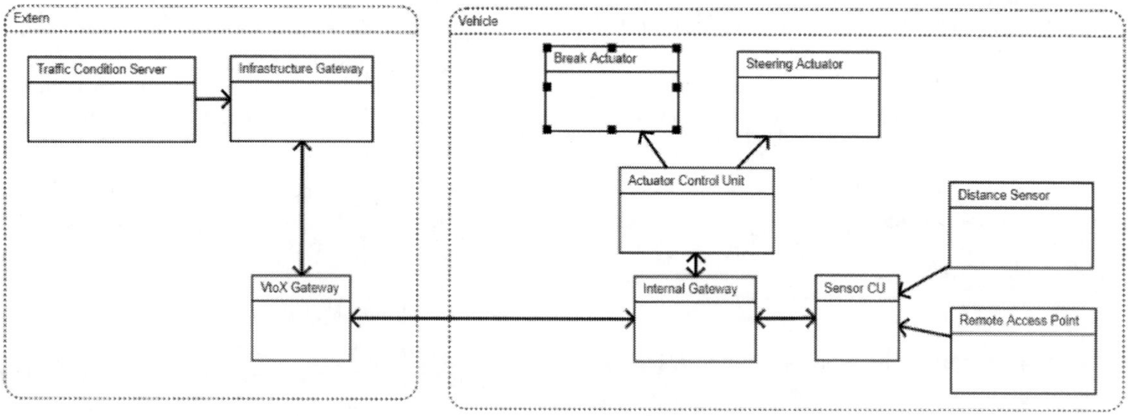

Figure 5. Automotive vehicle example model using FMVEA

to meet the actual security level needed. Future work will include the following points:

- Improve the threat database of ThreatGet.
- Integrate additional security countermeasures with MORETO's database like as ISO 27000 for information security management systems.
- Improve the risk evaluation methodology to be suitable for complex models.
- Enhance the OnSecta building blocks to manage more characteristics and relationships of threats, vulnerabilities, and security countermeasures.

Acknowledgment

The research project "Cybersicherheit für Verkehrsinfrastruktur- und Straßenbetreiber" (CySiVuS, in English: "Cybersecurity for transport infrastructure and road operators") (Project-Nr. 865081) is supported and partially funded by the Austrian National Security Research Program KIRAS (Federal Ministry for Transport, Innovation

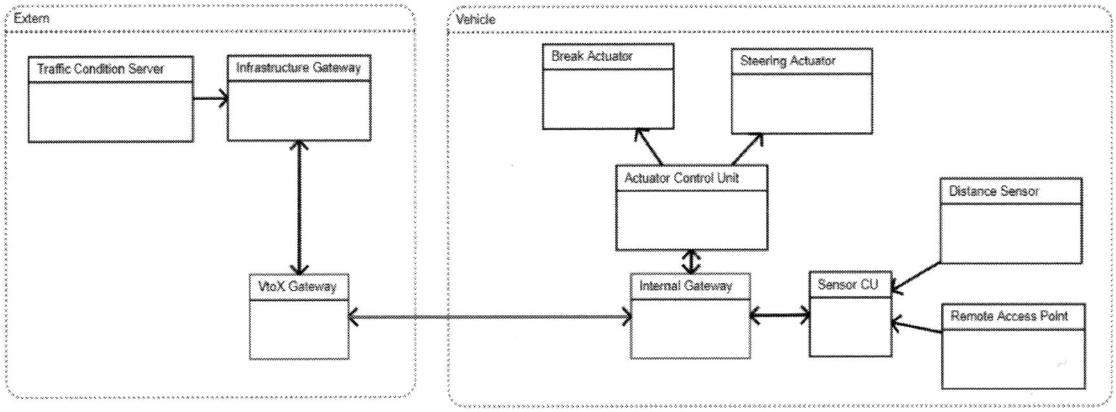

Figure 6. Affected elements and connections regarding rule the applied rules

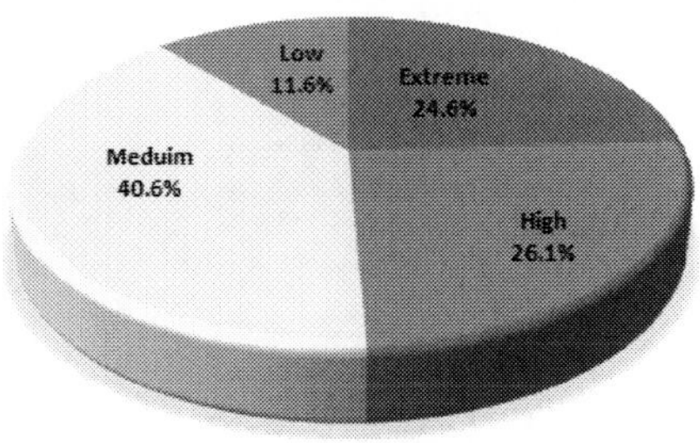

Figure 7. Statistical percentage of risk levels

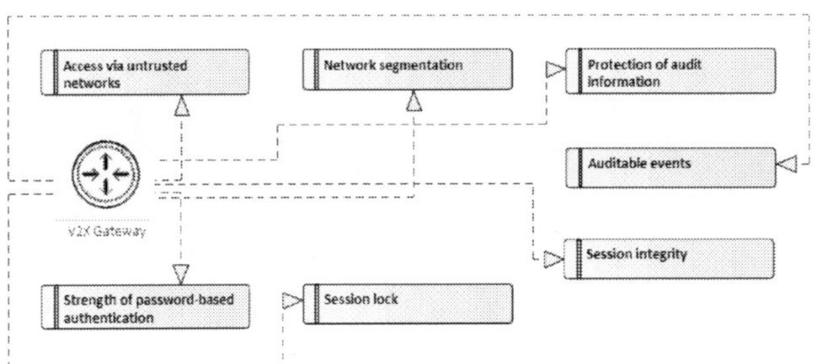

Figure 8. List of IEC 62443-4-2 security requirements of the V2X Gateway

and Technology (BMVIT) and Austrian Research Promotion Agency (FFG) 2017).

References

[1] Netscribe Blog, "Applications of IoT in the automotive industry | netscribes blog." [Online]. Available: https://www.netscribes.com/the-present-and-future-role-of-automotive-iot/ [accessed on: 2019-05-19].

[2] Carrie Cox and Andrew Hart, "How autonomous vehicles could relieve or worsen traffic congestion," Here Technologies, Tech. Rep., 2017.

[3] C. Schmittner, M. Latzenhofer, A. M. Shaaban, and M. Hofer, "A proposal for a comprehensive automotive cybersecurity reference architecture," in VEHICULAR 2018, The Seventh International Conference on Advances in Vehicular Systems, Technologies and Applications, 2018.

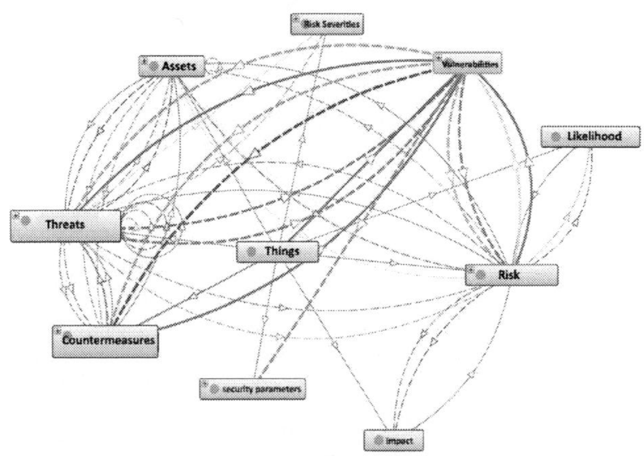

Figure 9. The ontological representation of threats, vulnerabilities, and security requirements

[4] A. Birni and T. van Roermund, "A multi-layer vehicle security framework," BU Automotive, NXP Semiconductors, Automotive Security, 2016.

[5] P. WARRENDALE, "SAE international releases updated visual chart for its "Levels of Driving Automation" standard for self-driving vehicles," 2018. [Online]. Available: https://www.sae.org/news/press-room/2018/12/sae-international-releases-updated-visual-chart-for-its-%E2%80%9Clevels-of-driving-automation%E2%80%9D-standard-for-self-driving-vehicles [accessed on: 2019-05-10].

[6] Electrek, "Tesla autopilot," 2019. [Online]. Available: https://electrek.co/guides/tesla-autopilot/ [accessed on: 2019.05.18].

[7] L. Chen and C. Englund, "Cooperative its - eu standards to accelerate cooperative mobility," in 2014 International Conference on Connected Vehicles and Expo (ICCVE). IEEE, 2014, pp. 681–686.

[8] ISO, "Iso 17427-1:2018 intelligent transport systems – cooperative its – part 1: Roles and responsibilities in the context of co-operative its architecture(s)," International Organization for Standardization, Standard, 2018.

[9] A. M. Shaaban, C. Schmittner, T. Gruber, A. B. Mohamed, G. Quirchmayr, and E. Schikuta, "Cloudwot-a reference model for knowledge-based iot solutions," in Proceedings of the 20th International Conference on Information Integration and Web-based Applications & Services. ACM, 2018, pp. 272–281.

[10] Wolf, Marko and Daly, PW, Security Engineering for Vehicular IT Systems, may 2009 ed. Vieweg and Teubner, 2009.

[11] M. Wolf and M. Scheibel, "A systematic approach to a qualified security risk analysis for vehicular it systems," Automotive-Safety & Security 2012, 2012.

[12] S. Strobl, D. Hofbauer, C. Schmittner, S. Maksuti, M. Tauber, and J. Delsing, "Connected cars - threats, vulnerabilities and their impact," in 2018 IEEE Industrial Cyber-Physical Systems (ICPS). IEEE, 2018, pp. 375–380.

[13] C. Schmittner, G. Griessnig, and Z. Ma, "Status of the development of isosae 21434," in European Conference on Software Process Improvement. Springer, 2018, pp. 504–513.

[14] KIRAS Security Research, "Cyber security for transport infrastructure- and road operators (cysivus)." [Online]. Available: https://www.kiras.at/en/financed-proposals/detail/d/cyber-security-for-transport-infrastructure-and-road-operators-cysivus/ [accessed on: 2019.05.02].

[15] IEC 62443-3-3, "Industrial communication networks - network and system security - part 3-3: System requirements and security levels," International Electrotechnical Commission, Tech. Rep., 2013.

[16] S. Mehar, S. M. Senouci, A. Kies, and M. M. Zoulikha, "An optimized roadside units (rsu) placement for delay-sensitive applications in vehicular networks," in 2015 12th Annual IEEE Consumer Communications and Networking Conference (CCNC). IEEE, 2015, pp. 121–127.

[17] ISO-SAE 21434, "Road vehicles – cybersecurity engineering cd," 2019. [Online]. Available: https://www.iso.org/standard/70918.html [accessed on: 2019.05.16].

[18] Oracle Asset Identification, "Overview to Asset Identification," 2013. [Online]. Available: https://docs.oracle.com/cd/E26228_01/doc.93/e21539/ [accessed on: 2019.04.28].

[19] Sparx Services CE, "Threat Modeling with STRIDE," 2019. [Online]. Available: https://cybersecurity.sparxservices.eu [accessed on: 2019.05.21].

[20] Z. Ma and C. Schmittner, "Threat modeling for automotive security analysis," Advanced Science and Technology Letters, vol. 139, 2016, pp. 333–339.

[21] C. Lévy-Bencheton, L. Marinos, R. Mattioli, T. King, C. Dietzel, and J. Stumpf, "Threat landscape and good practice guide for internet infrastructure," EU Agency for Network and Information Security (ENISA), 2015.

[22] C. Miller and C. Valasek, "A survey of remote automotive attack surfaces," black hat USA, vol. 2014, 2014, p. 94.

[23] A. Shostack, "Experiences threat modeling at microsoft," vol. 413, 2008.

[24] C. Schmittner, Z. Ma, and P. Smith, "Fmvea for safety and security analysis of intelligent and cooperative vehicles," in International Conference on Computer Safety, Reliability, and Security. Springer, 2014, pp. 282–288.

[25] I. E. Commission et al., "Iec 60812: Analysis techniques for system reliability-procedure for failure mode and effects analysis (fmea)," Geneva, Switzerland: International Electrotechnical Commission, 2006, pp. 1–93.

[26] C. Schmittner, Z. Ma, E. Schoitsch, and T. Gruber, "A case study of fmvea and chassis as safety and security co-analysis method for automotive cyber-physical systems," in Proceedings of the 1st ACM Workshop on Cyber-Physical System Security. ACM, 2015, pp. 69–80.

[27] C. Schmittner, Z. Ma, C. Reyes, O. Dillinger, and P. Puschner, "Using sae j3061 for automotive security requirement engineering," in International Conference on Computer Safety, Reliability, and Security. Springer, 2016, pp. 157–170.

[28] A. M. Shaaban, E. Kristen, and C. Schmittner, "Application of iec 62443 for iot components," in International Conference on Computer Safety, Reliability, and Security. Springer, 2018, pp. 214–223.

[29] Automotive Sensors and Electronics Expo 2017, "Automotive sensors and electronics expo 2017," 4th Annual Conference and Exhibition, 2017. [Online]. Available: http://www.automotivesensors2017.com [accessed on: 2019.04.27].

[30] embitel, "ECU is a three letter answer for all the innovative features in your car: Know how the story unfolded," 2017. [Online]. Available: https://www.embitel.com/blog/embedded-blog/automotive-control-units-development-innovations-mechanical-to-electronics [accessed on: 2019.05.15].

[31] IEC 62443-3-2, "IEC 62443 security for industrial automation and control systems - part 3-2: Security risk assessment and system design." ISA, Security Standard Committee Draft for Vote (CDV) IEC 62443-3-2 ED1.

My Connected Car Is Under Attack: "TPM" , "TPM" HELP ME

Jeevan Visvesha
ANI technologies private limited
Bangalore, India
jeevan.visvesha@olacabs.com

Abstract— **Raising trend in connecting vehicle prompts us to answer how secure they are. They are no less than a massive computer. There are around sixty to eighty Electronic Control Unit (ECU) in the vehicle. Modern automotive technologies are required to enhance the consumer in-car experience which at the same time expands the attack surface, opening up a host of new vulnerabilities. Presence of around eighty ECUs additionally mandates for secure in-vehicle communication between the vital ECUs, that is not existent in present vehicles. Telematics Control unit (TCU) is commonly utilized in vehicles to act as sort of a gate in uplinking any vehicle information to Infrastructure and downlinking files from Infrastructure to vehicle. There are a lot of complexities involved with building connected system. Solutions for a portion of the complexities present us with imperative issues like identification and authentication, which may be of high impact. The paper can provide answers to following queries: How TPM helps in moving towards a secure authentication with the infrastructure? How TPM helps in securing the Uplinking and OTA communication? How ECU to TPM communication can be secured?**

Keywords- TPM; V2I; OTA; Secure protocol; Authentication

I. Introduction

Evolution from a traditional vehicle to connected vehicle has introduced crucial changes. The Original Equipment Manufacturer (OEM) has invested huge amounts of money on connected vehicles to boost the client comfort in driving, diagnosing, prognostic maintenance, driver assisted systems, drive patterns, vehicle tracking systems, automated controls and different luxury functionalities. Beside plain edges of connected vehicles, it exposes for many attack surfaces. However, the industry failed in predicting, understanding and addressing to the security dangers and vulnerabilities identified with associated vehicle.

As a result of lack of security in existing vehicle protocol, ton of attacks is found from sniffing the messages to flashing the malicious software configuration file, thereby affecting the critical features of the vehicle. The effect of which can be as little as sniffing sensitive data spillage to as large as human life harm. This vulnerability becomes abundant larger whenever a new device is added because of the lack of robust device identification and authentication.

This paper proposes the thought of security through Trusted Platform Module (TPM) to give resistance against the attack vectors and recognize and authenticate the ECUs securely [12][13].

The TPM that is integral part of the solution is a kind of Hardware Security Module (HSM). It is a worldwide standard for a Secure crypto processor, a devoted microcontroller intended to secure hardware through integrated cryptographic keys [8].

The paper consists of following four sections. Section II portrays the issue existing in the present usage. Section III displays the solution to mitigate the attack surfaces. Section IV provides the conclusion and future work.

II. Problems in Existing Connected vehicle

This section quickly portrays on the issues that are existing in the current connected vehicle usage. These issues result in a huge harm.

A. Device Identification

Identifying any TCU in a connected vehicle plays a significant role in processing the received information and taking an applicable call on functionalities. However, in present connected vehicle scenarios, identification of TCU is occurring on entities that are susceptible to sniffing and counterfeiting.

When exchanging sensitive information or issuing some software configuration file to device, it is constantly important to recognize a specific device. Presently in several TCUs, the identifiers which are utilized are International Mobile Equipment Identity (IMEI), printed sequential number on device, which can be easily read. TCU vendors additionally will in general utilize only a gradual sequential number which can be effectively anticipated. Similar identifiers can be utilized in counterfeited TCUs so as to act like an authentic device as mentioned in [1][2][7] on counterfeiting electronic components.

B. Device Authentication

The Authenticity of a device plays a significant role, because it is intended to validate its identity and only authorized devices are connecting to the network securely. Present TCU manufactures are accustomed to simple authentication mechanism to ease the method and reduce processing complexity.

The present generally utilized strategy for authentication varies from simple token based or user name and password to relatively secure X.509 certificate based authentication. These authentication strategies are sufficiently dependable as long as this sensitive information's are stored in a secured memory in the Server and device.

C. Firmware/Software update or Configuration file change

Firmware Over The Air (FOTA)/Software Over The Air (SOTA) is an approach used to update software/firmware of any ECU over the air. This feature helps in upgrading or performing diagnostics without taking the vehicle to the Service station. Any malicious software update result in a huge devastation to the system in the vehicle. It very well may be as basic as change of vehicle parameters to complex as remotely accessing or controlling the vehicle mentioned in [3][4][10].

The current industry updating or providing configuration push over the air is actualized on Cyclic Redundancy Check (CRC)/checksum mechanism. An assailant can make his own malicious file with appropriate determined CRC/checksum. There is no validation on the device whether or not the configuration files are from the trustworthy server, that has to be addressed in an exceedingly secure method

D. Probable implementation mistakes in correct usage of TPM

We may have seen solely advantage and secure part of using TPM. Are there any Attack areas for the TPM? Indeed, we do have if TPM is not utilized in the right manner.

Regardless of the reality, we know the active attack on TPM is possible, yet hard to perform and requires expensive devices. A passive attack is still possible with modest and simple strategies.

We realize that TCU microcontroller is connected to TPM through I2C/SPI lines. As these TPM cannot do bulk encryption, these secret keys should be shared to microcontroller when required. These lines can be sniffed to fetch the sensitive data and keys [5][7].

III. MITIGATION

This section clarifies the proposed flow for mitigating issues which was referenced in the previous segment. Our plan is to principally touch upon the secure integration of TPM to the prevailing ECU

A. Device identification

- Each TPM is a unique. Master ID will be burnt during the manufacturing process which cannot be read or altered by anyone
- Changing the master ID is equivalent to changing into new device.
- Keeping this Master ID as seed, a key pair is generated known as EK_pub and EK_priv (EK = Endorsement Key) [12]
- Another set of Key is generated after ownership is claimed on TPM. This is Storage Root Key (SRK). using this, SRK_pub and SRK_priv is generated.
- EK is specific to device and SRK is specific to owner
- Microcontroller will have its own manufacture ID (μC ID)
- Secure uploading of all μC ID, EK_pub and SRK_pub is done to the server which is referred as inventory list as shown in Figure 1.

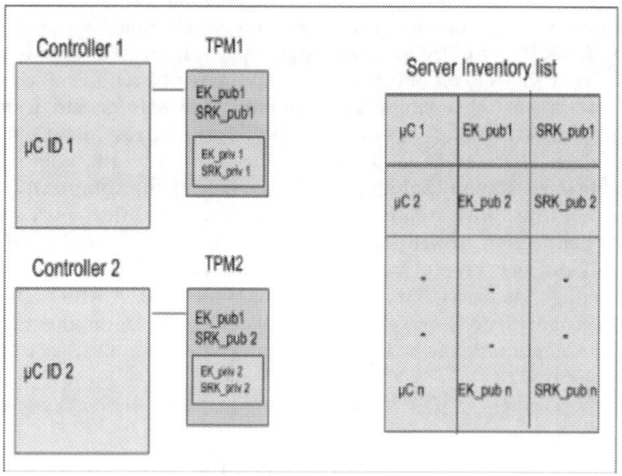

Figure 1. Device identification

B. Device Authentication

Device authentication is the most important mechanism in any of the connected world, where the server wants to identify the device and also device wants to identify the server, thereby mitigating the two entities impersonating each other.

a) Initiation step
   - Device with a TPM first connects to the server and requests to initiate for authentication
   - Server_pub key is pre-stored in TCU during provisioning process.
   - The device shall generate a Random number known as "Nonce".
   - The Nonce and μc ID are encrypted and sent to the server as shown in Figure 2. $\text{Encrypt}_{\text{server\_pub}}$[Nonce + μC ID]

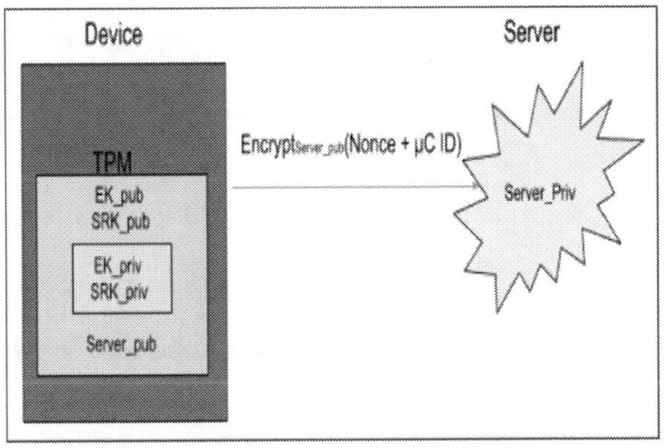

Figure 2. Initiation flow

b) Challenge Flow

- ❖ The server now decrypts the received value using Server_priv key.
- ❖ Once it decrypts, the Server will generate the Auth Validator (a random string) shown in Figure 3.1.
- ❖ Using the µC ID, it fetches the mapping of EK_pub and SRK_pub from the server inventory list.
- ❖ Now the Nonce + Auth Validator is signed using Server_priv and encrypted 1st using EK_priv and then using SRK_priv.
- ❖ This data is sent to the device as shown in Figure 3.2

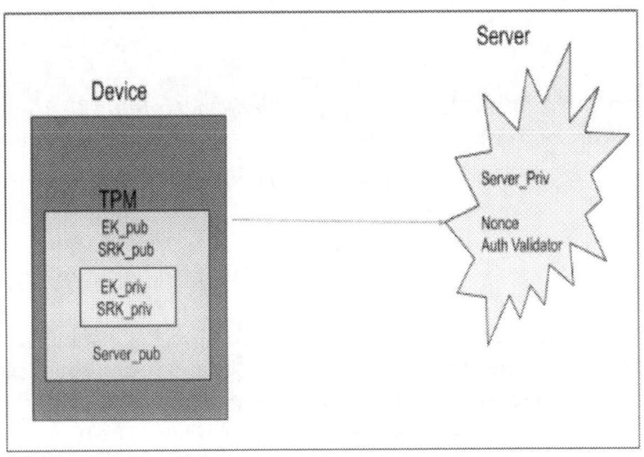

Figure 3.1. Challenge Flow

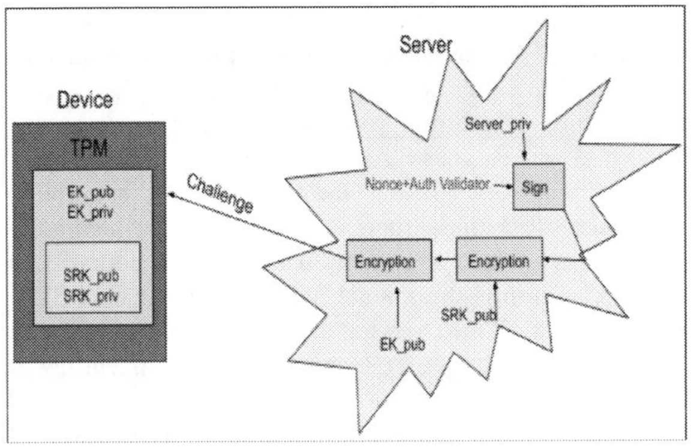

Figure 3.2. Challenge Flow

c) Verification flow

- ❖ The encrypted nonce is decrypted first using EK_priv and then with SRK_priv to prove the ownership.
- ❖ The signed value is verified using Server_pub key to authenticate whether it has come from a trusted server.
- ❖ Once signature is verified, The obtained Nonce is verified with the nonce which was generated during the initialisation flow.
- ❖ Once Nonce matches, the Auth validator is stored in the TPM secure memory as shown in Figure 4.

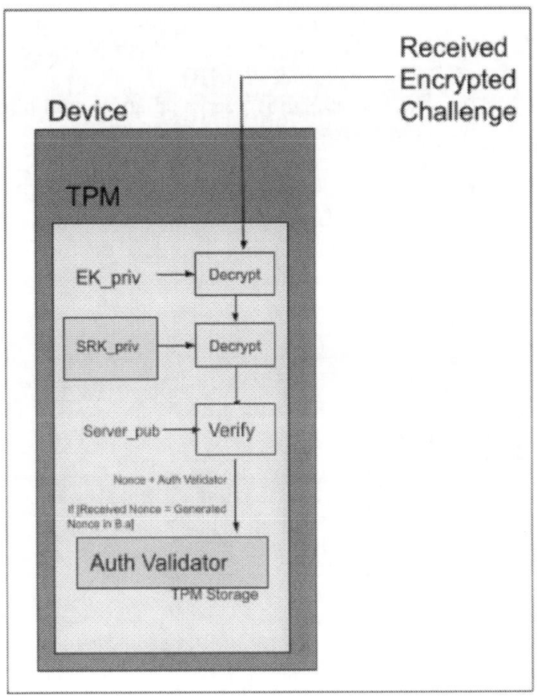

Figure 4. Verification Flow

As a result of the usage of Nonce, replay attack can be avoided which is the real security concern in the present industry. Auth Validator is generated by the server and same Auth Validator is shared securely to TPM. This **Auth Validator** will serve as a symmetric key for encrypting further data to the cloud, thereby solving the confidentiality of the information exchanging mutually. This Process of step a, b and c performs the mutual authentication where TCU can authenticate the server securely and server can authenticate and trust the TCU, which is non-existent in the automotive domain. Introduction of robust security mechanisms like this can provide a **new dimension of security** for the connected vehicle.

C. Firmware/Software update or Configuration file change

Any software update and configuration change must be performed securely, as the dangers were clearly referenced in the previous section. Integration of TPM with the TCU helps in doing this task securely. The serious issue in any of the crypto process is storing of the sensitive keys which are utilized for performing signing, encryption. Storing of these sensitive keys in a secure memory is a major task. In any case, TPM also provides a secure storage highlighting feature to store sensitive keys.

a) Code Signing process

- The Server will generate Server_priv key and store it securely
- The Server_pub key is encrypted and shared to TCU like how it shared Auth Validator in Device Authentication [B].
- The Server_pub is decrypted and stored in the TPM Secure memory.
- EK_pub is already available with the Server
- Now the Software or configuration file is signed using Server_priv key and encrypt the signed packet using EK_Pub of respective device and share to the device as show in Figure 5.

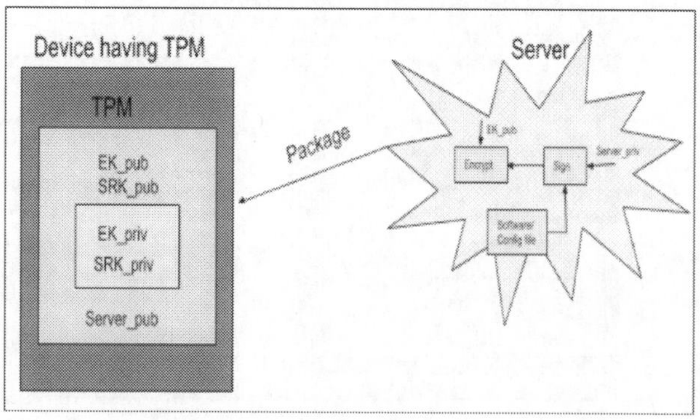

Figure 5. Code signing process

b) Verification process

- The received package is protected as it is encrypted which solves the confidentiality issue.
- This package is decrypted using device EK_priv.
- The decrypted package is verified using Stored Server_pub key. This verifies the integrity and also verifies it is from a trusted server as shown in Figure 6.

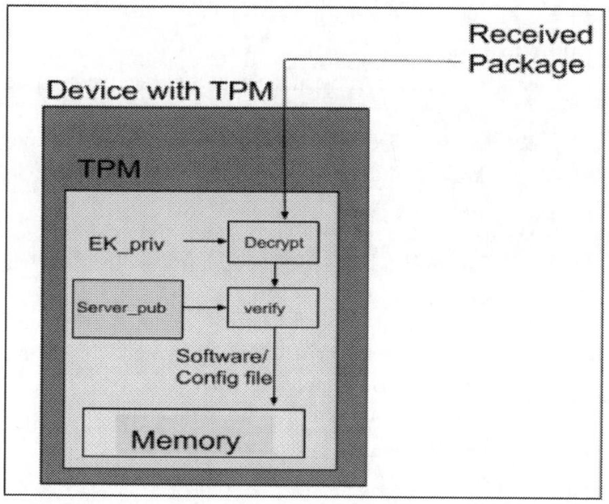

Figure 6. Verification of Package

D. Probable solution to mitigate implementation mistakes while marrying TPM to ECU

As described in the previous section, TPM can add to powerful security modules. However, if not implemented and incorporated in the correct manner, it is prone to several security risks. Below are the considerations to be taken care while marrying TPM to any critical ECU.

- Use the TPM, which has a BGA package (pins are underneath and hard to find the pins for probing). Prefer not to use packages with exposing leads like QFN/TSSOP packages.
- The I2C/SPI lines communicating with the microcontroller must be routed through the internal layers while designing Printed Circuit Board (PCB). So making the communication lines difficult for probing as mentioned in [6].

IV. Conclusion and Future Work

The primary focus of this paper was to provide efficient and practical solutions for some of critical threats in the connected vehicle environment, which was explained in Section II. These attack surfaces can be segregated into 2 major things, one is a physical attack and another is a remote attack. There are several researches done on the technical aspects of the security of connected cars. This study combines the existing research on the technical security aspects of connected vehicles along with the improvisation of security in connected vehicles. Usage of TPM is already proven in networking domains in enhancing the security. Recently automotive domain started using it. It is worth looking at matured domains to borrow certain technology to empower security posture in ever growing automotive world. Another research question that can be examined and is not covered in this paper is about the V2V secure communication. How do we extend this connected vehicle concept to prevent accident or enhance the safety of passenger or driver by connecting to nearby vehicle when in danger? Thus, with the current and growing awareness of the importance of hardware security, trustworthy connected vehicle systems can be deployed in the coming years.

References

[1] J. Laidlaw, "Counterfeit hardware may lead to malware and failure", HACKADAY, 18 June 2019.

[2] C.S Jeena, "Counterfeit threat for electronic industry on rise: plug it", The Holography times, Volume 8, Issue 24, 2014.

[3] S. Nie, L. Liu, W. Zhang and Y. Du, "Over-the-air: How we remotely compromised the Gateway, BCM and Autopilot ECUs of Tesla cars", Blackhat USA, August 2018.

[4] "Car Hacking Research: Remote attack Tesla Motors", Keen Security lab of Tencent, 19 September 2016.

[5] J. Winter and K. Dietrich, "A hijacker's guide to communication interfaces of the trusted platform module", ScienceDirect, Volume 65, March 2013.

[6] "AN928.1 EPR32 Series 1 Layout Design guide", Silicon labs

[7] P. Papadimitratos, L. Buttyan, T. Holczer, E. Schoch, J. Freudiger, M. Raya Z. Ma, F. Kargl, A. Kung and J.P Hubaux, "Secure vehicular Communication Systems: Design and Architecture", IEEE Communications Magazine, 30 December 2009.

[8] F. Kargl, P. Papadimitratos, L. Buttyan , M. Muter, E. Schoch , B. Wiedersheim T.V. Thong, G. Calandriell, A. Held and A. Kung, "Secure Vehicular Communication Systems: Implementation, Performance and research Challenges", IEEE Communications Magazine, 25 November 2008.

[9] "Secure Vehicular Communications: Security Architecture and Mechanisms for V2V/V2I", SeVeCom, Delivarable 2.1, 2007- 2008.

[10] S. Nie, L. Liu and Yuefeng Du, "Free-fall: hacking tesla from wireless to CAN bus", blackhat, September 2016.

[11] T. Leinmuller, "Car2x Communication - Challenges, standardization and implementation in Europe and in the US", 2007.

[12] J.S. Suresh and L. Jongkun, "A TPM based architecture for Secure VANET", Indian Journal of Science and Technology, July 2015.

[13] G. Guette and O. Heen. "A TPM based architecture for improved security and anonymity in Vehicular Ad-hoc Networks", IEEE Vehicular Networking Conference, 2009.

[14] "TPM 2.0 Library Specification", Trusted Computing Group, 29 September 2016.

[15] I. Foster, A. Prudhomme, K. Koscher, and S. Savage, "Fast and Vulnerable : A story of Telematics Failures", USENIX, 2015.

**International Academy, Research,
and Industry Association (IARIA)**
412 Derby Way
Wilmington, DE 19810

ISBN 978-1-5108-9047-3

9 781510 890473

CURRAN ASSOCIATES INC.
proceedings
.com

International Workshop on Quantum Chromodynamics – Theory and Experiment (QCD@Work 2018)

EPJ Web of Conferences Volume 192 (2018)

Matera, Italy
25 - 28 June 2018

edp sciences